HETEROGENEOUS
CATALYSIS

PRENTICE HALL INTERNATIONAL SERIES
IN THE PHYSICAL AND CHEMICAL ENGINEERING SCIENCES

NEAL R. AMUNDSON, SERIES EDITOR, *University of Houston*

ADVISORY EDITORS

ANDREAS ACRIVOS, *Stanford University*
JOHN DAHLER, *University of Minnesota*
THOMAS J. HANRATTY, *University of Illinois*
JOHN M. PRAUSNITZ, *University of California*
L. E. SCRIVEN, *University of Minnesota*

AMUNDSON *Mathematical Methods in Chemical Engineering: Matrices and Their Applications*
BALZHIZER, SAMUELS, AND ELLIASSEN *Chemical Engineering Thermodynamics*
BRIAN *Staged Cascades in Chemical Processing*
BUTT *Reaction Kinetics and Reactor Design*
DENN *Process Fluid Mechanics*
FOGLER *Elements of Chemical Reaction Engineering*
HIMMELBLAU *Basic Principles and Calculations in Chemical Engineering, 5th Edition*
HOLLAND *Fundamentals and Modeling of Separation Processes: Absorption, Distillation, Evaporation, and Extraction*
HOLLAND AND ANTHONY *Fundamentals of Chemical Reaction Engineering*
KUBICEK AND HAVACEK *Numerical Solution of Nonlinear Boundary Value Problems with Applications*
LEVICH *Physiochemical Hydrodynamics*
MODELL AND REID *Thermodynamics and its Applications, 2nd edition*
MYERS AND SEIDER *Introduction to Chemical Engineering and Computer Calculations*
NEWMAN *Electrochemical Systems*
PRAUSNITZ, LICHTENTHALER, AND DE AZEVEDO *Molecular Thermodynamics of Fluid-Phase Equilibria, 2nd edition*
PRAUSNITZ ET AL. *Computer Calculations for Multicomponent Vapor-Liquid and Liquid-Liquid Equilibria*
RHEE ET AL. *First-Order Partial Differential Equations: Theory and Applications of Single Equations. Volumes I and II*
RUDD ET AL. *Process Synthesis*
SCHULTZ *Diffraction for Materials Scientists*
SCHULTZ *Polymer Materials Science*
VILLADSEN AND MICHELSEN *Solution of Differential Equation by Polynomial Approximation*
WHITE *Heterogeneous Catalysis*
WILLIAMS *Polymer Science Engineering*

HETEROGENEOUS CATALYSIS

MARK G. WHITE

School of Chemical Engineering
Georgia Institute of Technology

PRENTICE HALL
Englewood Cliffs, New Jersey 07632

o 382-0944

Library of Congress Cataloging-in-Publication Data

WHITE, M. G. (MARK G.),
  Heterogeneous catalysis.

  (Prentice-Hall international series in the physical
and chemical engineering sciences)
  Includes bibliographies and index.
  1. Heterogeneous catalysis. I. Title. II. Series.
QD505.W47 1989   541.3'95   88-32541
ISBN 0-13-387739-6

Editorial/production supervision: BARBARA MARTTINE
Manufacturing buyer: MARY ANN GLORIANDE

 © 1990 by Prentice-Hall, Inc.
A Division of Simon & Schuster
Englewood Cliffs, New Jersey 07632

Printed in the United States of America

10 9 8 7 6 5 4 3 2 1

ISBN 0-13-387739-6

PRENTICE-HALL INTERNATIONAL (UK) LIMITED, *London*
PRENTICE-HALL OF AUSTRALIA PTY. LIMITED, *Sydney*
PRENTICE-HALL CANADA INC., *Toronto*
PRENTICE-HALL HISPANOAMERICANA, S.A., *Mexico*
PRENTICE-HALL OF INDIA PRIVATE LIMITED, *New Delhi*
PRENTICE-HALL OF JAPAN, INC., *Tokyo*
SIMON & SCHUSTER ASIA PTE. LTD., *Singapore*
EDITORA PRENTICE-HALL DO BRASIL, LTDA., *Rio de Janeiro*

This text is dedicated to honor the memories of
O. A. Gilmore and E. W. White.

CONTENTS

PREFACE

This textbook develops simple principles of heterogeneous catalysis from the chemistry of homogeneous reactions and the physics of fluid/solid interactions. We illustrate principles of catalysis for a few simple systems by which the student may then understand complicated systems. As such, the text is a teaching tool rather than an exhaustive reference book. In several instances we present simple but descriptive discussions of topics rather than rigorous expositions.

The goal of this manuscript is to provide a bridge between the chemistry knowledge generally held by third-year university students majoring in physics, chemistry or chemical engineering and that needed to understand simple problems in catalysis. In our experience, part of the information needed to bridge this knowledge gap was developed in formal courses (quantum and statistical mechanics, advanced thermodynamics, solid-state, inorganic, and organometallic chemistry). The remainder of the bridge-building information was developed through experience in the catalysis laboratory. Since it is impossible to include in one textbook all the information in these formal courses and to deliver all the experience learned in the laboratory, we choose only that information which proved most profitable to the students enrolled in our graduate catalysis course.

We have used this manuscript in teaching a graduate-level course titled "Heterogeneous Catalysis" for the past six years. Although most of the students enrolled in the course were chemical engineers, a few physicists and chemists have also taken the course. Most of the students learned the material quickly, irrespective of their preparation prior to enrolling in the class. We recommend that the students, scientists and engineers using this book have a knowledge of chemistry through physical chemistry.

The text is written in three parts: Sorption Physics at Fluid/Solid Interfaces, Experimental Methods, and Reaction Kinetics. In the first part, a microscopic picture of fluid/solid interfaces is developed through a series of increasingly complex models of the surface. From these surface models and from the principles of intermolecular potential energy functions, we illustrate the surface/sorbate interactions that lead to the calculation of adsorption energies from potential energy curves. From these microscopic descriptions of sorbate/sorbent interactions, we move to macroscopic descriptions of the same. Here, we use equilibrium thermodynamics to calculate adsorption enthalpies at constant spreading pressure. Next, the Langmuir isotherm is developed from two different models. The first model assumes that a dynamic equilibrium is established between the phases with the frequency of surface encounters described by the collision theory derived from the kinetic theory of gases. The other model is based on statistical mechanics and the sorption phenomenon is modeled as an ensemble of microstates each having the same energy. Following these theoretical considerations, we present experimental methods to develop these theories and to describe the surface morphology and reactivity.

Part II begins with a discussion of the classical tools of surface characterization:

total and selective sorptions in volumetric devices. Next, we discuss techniques to classify the surface according to sorption energy using static and dynamic techniques. Infrared spectroscopy is introduced as a means to describe the vibrational states of chemisorbed probe molecules and reactive intermediates. Following these classical tools, we discuss briefly four ultrahigh vacuum techniques and how they are used to characterize the surface. Part II ends with a discussion of experimental reactors. We focus on the relative merits of reactor types and describe the virtues of each towards the goal of characterizing reactivity of heterogeneous catalysts. After the merits of the laboratory reactors are discussed, we briefly describe the diagnostics for transport disguise mechanisms, how these mechanisms affect the observed kinetics and how they can be eliminated by experimental design and technique.

Part III begins with the fundamentals of homogeneous reaction kinetics of "unimolecular" decompositions. With the eventual discussion of the Rice, Ramsperger, Kassel, and Marcus (RRKM) theory, we discuss the principles underlying all bimolecular, gas-phase reactions which set up the future discussion of reactions at gas/solid interfaces. Next, we show how the phenomenological rate laws for homogeneous reactions are simplifications of the more exact ones developed previously. We show how data of conversion versus reaction time in a batch reactor may be fit to determine the parameters of the integrated rate laws. This discussion of fitting data to homogeneous rate laws is followed by a similar discourse for heterogeneous rate laws. We introduce the general rate laws, such as Langmuir-Hinshelwood and Rideal-Eley, followed by specialized rate equations such as Temkin-Pyzhev and Mars-van Krevelen. This discussion on specialized rate laws is followed by case studies in which we describe how data are fit to complicated rate equations. We close the discussion of rate equations with a brief discourse on the compensation effect as applied to heterogeneous catalysts. Part III ends with a discussion of isotopically labeled compound studies. We describe two types of experiment: static and kinetic. This is followed by a discussion on how isotopic compounds are used together with kinetic studies to establish a reaction mechanism. We include brief reviews of only three types of labeled compounds: deuterium-labeled hydrocarbons, isotopic carbon compounds, and labeled oxygen.

We have included enough material in this text to fill a 16-week semester course. For a semester course, one could proceed through the book in the sequence of chapters as written. Should the instructor decide to vary the presentation sequence, we suggest beginning with the first half of Chapter 1. For a ten-week quarter course, we omit the following parts of the text: last half of Chapter 1 (quantum mechanics), Chapter 2, the sections of Chapter 7 describing RRK, RRKM and Transition State theories, and the sections of Chapter 8 describing Isotopic Exchange Reactions. For the case where reactor design, kinetics, and catalysis are taught in the same course, this text could be used for the kinetics and catalysis part of the course. This text does not develop reactor design principles, and transport effects are discussed only as they disguise the intrinsic kinetics.

We gratefully acknowledge the financial support from the Georgia Tech Foundation and the release time provided by the School of Chemical Engineering. We thank Professor Gary W. Poehlein, Associate Vice President of Research (Georgia Tech Research Institute) for his continued support and encouragement. We must acknowledge the help of Dr. R. K. Beckler and Dr. D. J. Rosenthal for the many long hours they expended at various computer terminals to produce this manuscript. Finally, thanks to the students who enrolled in the graduate heterogeneous catalysis course for their help in revising the manuscript.

Mark G. White

HETEROGENEOUS
CATALYSIS

PART I

SORPTION PHYSICS AT FLUID/SOLID INTERFACES

The sorption of gases onto solids is a complicated phenomenon depending upon the characteristics of the fluid and solid, in the most general case. We shall discuss here the physical picture of the well-defined interface and the models to describe the same. An attempt is made to qualitatively discriminate between chemisorption and physisorption. Next, the molecular description of sorbate/sorbent interaction will be presented from the viewpoints of potential energy diagrams and quantum mechanics. This molecular description is followed by a macroscopic picture of such interactions as defined by Gibbs adsorption thermodynamics. A lengthy discussion follows of sorption isotherms developed from classical collision theory and from statistical thermodynamics. Correlations of some real isotherms are summarized, and the applications are discussed. Finally the kinetics of adsorption/desorption processes are presented.

CHAPTER 1

MICROSCOPIC DESCRIPTION

OF SORBATE/SORBENT INTERACTIONS

Our experience with adsorbents usually involves only the process application, with little thought given to the actual events that are ensuing. We know that molecular sieves are excellent agents for separating hydrocarbon streams from water and CO_2 and that silica gel is an excellent drying agent; and so on. These and other industrial adsorbents are high surface area solids showing very "complicated" surfaces. The surface not only has a complex physical texture but also exhibits sorption sites having a distribution of sorption energies undoubtedly arising from the surface texture and composition. The modeling of such industrial sorbents is indeed a challenging task; the best approach may be to establish models of such surfaces to account for only some of the more important features of the real sorbent.

The most simple fluid/solid interface is the plane boundary which is geometrically smooth to divide the fluid and the solid phases. No attractive forces are present between the fluid species and the solid. For some types of very simple sorptions showing little attraction between gas phase molecules and the solid, this idealization may be adequate (e.g., physisorption). Clearly the plane boundary is inadequate for all types of sorptions involving strong forces of attractions between gas and solid phases.

The attracting plane represents the first approximation of a real surface over the plane boundary. It is assumed that an attractive force normal to the plane exists everywhere as a 2-D continuum between the solid and the fluid phase. The most simple model of the normal forces assumes simple adhesion as the gaseous species encounters the plane. One could imagine this situation as molecular scale "fly-paper" where adhesion occurs only if the molecule encounters the surface boundary by collision. There are no forces of attraction away from the surface and as such there can be no vibrational spectrum of the surface/sorbate species "bond." With the adhesion model there can be only monolayer coverage. To include the aspects of surface normal species vibrations, which do exist and can be documented by various spectroscopy tools, and multilayer coverage, it is necessary to have surface attractive forces which vary as a function of distance from the plane. This model allows for the dynamic equilibrium of adsorption/desorption to exist for multilayer sorptions.

Both the plane boundary and the attracting plane models showed a surface without "texture"; however, experience shows such texture is present. In the most general sense the term "texture" may represent geometric variations in the surface height above some arbitrary mathematical interface or the "texture" may in fact represent the variation in the force field along the surface which could be flat in the geometric sense.

Imagine a crystalline solid say a metal cleaved along a low index plane, as given in Figure 1 - 1, which shows one imperfection, an impurity, in the first row of atoms. A potential, $-U$, may be described as existing in the space around each atom.

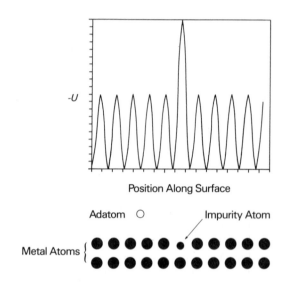

Figure 1 - 1 Model of Defected Surface and Its Potential

This potential generally depends upon electronic state of each individual atom *and* the arrangement of the atoms in the crystal. Imperfections in the crystal, such as vacancies, impurity atoms, and altervalent atoms (in oxides), will alter the potential function. We show here the qualitative effect on the general shape of the potential function for replacing a matrix atom by an impurity atom of a different size.

The effect of a vacancy in the solid is to change the shape of the surface potential function. If the impurity atom were removed from the surface, Figure 1 - 1 would represent the effect of geometric "texturing" upon the surface potential. The process of removing surface atoms at selected lattice points could be repeated at will

Chapter 1 Picture of the Fluid-Solid Interface

to model a highly "textured" surface which in fact would represent a polycrystalline solid showing a distribution of plane indices including both high and low index planes.

These models have assumed no effect of the gas upon the solid, but there is ample evidence to show such effects do exist. For example, single crystal Pt samples exposed to cycles of oxidizing/reducing atmospheres and examined by low energy electron diffraction (LEED) show the crystal surface to develop steps (or kinks) upon exposure to the O_2 and the steps coalesced by H_2 exposure.[1] Thus, it is possible to create high index surfaces out of low index, basal planes. Such a solid may be called a deformable solid; that is, the surface structure may deform and become restructured as a result of the adsorption event. The restructuring may or may not be reversible upon desorption.

The characterization of the solid/fluid interface has been confined to single crystals presenting a well-defined plane to the sorbing fluid. In the deformable solid, the sorption event caused restructuring of the surface to create some higher index planes (i.e., steps). These "steps" in the single crystal are but one example of surface *heterogeneity*. These surface imperfections, kinks, and steps are possible sites for chemisorption and reaction.

In the most general sense, a heterogeneous surface is characterized by a distribution of surface sites having different sorption energies. The result of such heterogeneity is an enthalpy of sorption which may vary with fractional coverage. The first sites to be covered liberate the highest heat of sorption; whereas the last sites to be covered release the lowest heat of sorption. Examples of some sorbents showing this heterogeneity include the polycrystalline solids (metal films, foils, and crystalline oxides), amorphous gels, and gels compressed into pellets. The genesis of this surface heterogeneity is a combination of the effects presented in the preceding discussion. Here we may see a grain of a crystal showing planes having highly stepped surfaces together with vacancies and defects. If the material is composed of pressed powder, then grain boundaries will generate even more imperfections. The picture which emerges is one of great complexity; the modeling of such a heterogenous surface from first principles is very difficult.

Chemisorption and Physisorption

There is a great temptation to characterize sorption processes as strong and weak sorptions so as to facilitate an in-depth study of each. However, at the very moment one begins to define the terms strong and weak sorptions, one is challenged to do so in such a manner as to evade all possible exceptions and examples. Nevertheless, we will present here a crude, qualitative discrimination between chemisorption (strong) and physisorption (weak), recognizing one may find examples of each which are contrary to the criteria in the list.

Table 1 - 1 Qualitative Distinction between Physisorption and Chemisorption

Criteria	Physisorption	Chemisorption
Forces of Attraction	van der Waals	Chemical Bonding
Heats of Adsorption (per mole)	2 - 6 kcal (8.4 - 25.1 kJ)	15 - 20 kcal (62.8 - 83.7 kJ)
Temperature Range of Sorption	Near saturation	Can occur at temperatures higher than saturation temp
Energy of Activation	None, nonactivated	May be activated
Selective Sorption Possible?	No	Yes
Multiple Layers of of Adsorbate Possible?	Yes	No

Theory of Adsorption Forces - Physical Adsorption

This discussion will focus on developing the theory of attraction and repelling forces between a fluid atom or molecule and atoms in the surface region of the solid. For physical adsorption, the attractive forces are those similar to nonideal gases: van der Waals. These forces are small for low density fluids; in the limit of zero density the forces are identically zero. For condensed phases, the van der Waals forces can be significant. The repulsion forces are those arising from the Pauli Exclusion Principle.

We limit the discussion to involve only atoms showing spherically symmetric electron clouds. Temporal fluctuations in the electron density about the nucleus give rise to the creation of instantaneous dipole moments. It is these instantaneous dipoles that create the very weak van der Waals attraction forces. As a first approximation, the fluctuations in the electron density may be modeled as a harmonic oscillator.[2a] The minimum potential energy of a pair of such oscillators, separated by a distance, r, will occur when the oscillations in the electron density are in phase. The total potential energy of these in-phase oscillators is *less* than that of the two at infinite separation; hence the forces will be attractive. These attractive forces have been called London Dispersion Forces after F. London. The potential energy of such dispersion forces is given by[2b]

$$U = -C/r^6 \qquad (1 - 1)$$

where the constant C is evaluated using London's theory.[2c]

$$C = (3/2)\alpha_1\alpha_2 I_1 I_2/(I_1 + I_2) \qquad (1 - 2)$$

and

α_1, α_2 = polarisability of atoms 1 and 2
I_1, I_2 = ionization potentials of atoms 1 and 2

Alternate expressions of the constant C have been developed;[3a, 3b, 3c]

$$C_{km} = 6mc^2\alpha_1\alpha_2/((\alpha_1/\chi_1) + (\alpha_2/\chi_2)) \qquad (1 - 3)$$

where

m = mass of electron
c = speed of light
χ = magnetic susceptibility

   More precise calculations of these van der Waals forces require proper accounting for higher order coupling of the variations in the electronic cloud; for example, dipole/quadrapole and quadrapole/quadrapole interactions. These higher order couplings are modeled as additive terms, c_1/r^8 and c_2/r^{10}, in the potential energy function.[3b] The contributions to the total potential energy, U, for these two terms may be as much as 10% and 1%, respectively.[3c]

   Repulsion forces cannot be developed so easily from first principles; usually these repelling forces are modeled as a potential energy of repulsion, U.

$$U_- = B_m r^{-m} \; ; \; m > 0 \qquad (1 - 4)$$

Values for m may be as high as 12. Therefore the potential energy between a pair of atoms separated by a distance, r, is described by

$$U_t = U_- + U_+ = B_{12} r^{-12} - Cr^{-6} \qquad (1 - 5)$$

The constant, B_{12}, may be eliminated from a consideration of the minimum condition at $r = r_o$; $(dU_t/dr)_o = 0$ so that

$$B_{12} = (1/2)Cr_o^6 \qquad (1 - 6)$$

Thus, the expression for the total pair potential energy is

$$U = C[0.5(r_o/r^2)^6 - (1/r)^6] \qquad (1 - 7)$$

Now, we turn our attention to the problem of one gas atom in the vicinity of an interacting plane. For the moment we assume the plane has no imperfections, steps, and so on, and that only the first layer of surface atoms will interact with the gas atom. This describes what is known as a van der Waals gas/van der Waals solid for adatom/surface interaction. A basic assumption of this theory is that the total potential energy for this adatom surface may be represented as a sum of pairwise contributions and that higher order contributions (two, three, four, and so on surface atoms/one gas atom) may be neglected. Thus, the total potential energy of this ensemble, U, is

$$U = \Sigma U_j = \Sigma[U_{-j} + U_{+j}] \qquad (1 - 8)$$

where U_{-j} are the repulsion potential energies of the jth pair and U_{+j} are the attraction potential energies of the jth pair. This sum must be taken over all possible surface/adatom pairs (j in number) and for most solids, j can be very large (1 cm^2 of plane surface will contain 10^{16} atoms, assuming an atomic diameter of 1 angstrom). This sum may be replaced by an integral over the dummy variable of integration, r, representing the distance between the adatom and the surface atoms:

$$U = \int_d^\infty [U_{+j} + U_{-j}]dr \qquad (1 - 9)$$

where d is the length of the normal from the surface to the adatom equal to twice the atomic radius of the adatom. For the attraction forces (U_+) between a surface atom/adatom, we have

$$U_{pair} = -C/r^6 \qquad (1 - 10)$$

In the total volume, V, of adatom gas there will be n adatoms such that

$$U(d) = \int(U_+)dV = \int(n/V)U_{pair}\, dV \qquad (1 - 11)$$

where n is the number of pairs on the surface, and dV is the differential volume enclosing the adatoms and surface atoms. Replacing dV by $(4\pi r^2)dr$ and using N to represent the number of adatoms per unit volume (n/V),

$$U_+(d) = -N\int_d^\infty (C/r^6)(4\pi r^2)dr = -4\pi NC/3d^3 \qquad (1 - 12)$$

Substituting $d = 2r_a$ yields (assuming the sum of the metal and adsorbate radii $= 2r_a$),

$$U_+ = -\pi NC/6r_a^3 \qquad (1 - 13)$$

where r_a = "radius" of adsorbate molecule at closest approach.[3d]

Consider now the repulsion term, U_r, using a general expression for the integrated potential energy:

$$U_r(r) = \{\pi NB/[(m - 3)(m - 2)]\}r_a^{(3-m)} \quad (1 - 14)$$

where m is a fitted variable. If we choose $m = 12$, then

$$U_t = -\pi NC/[6r_a^3] + \pi NB/[90r_a^9] \quad (1 - 15)$$

At the minimum of the total potential energy, $dU_t/dr = 0$, so

$$dU_t/dr = -\pi NC(-3)/[6d^4] - \pi NB/[10(d)^{10}] \quad (1 - 16)$$

or

$$B = 5Cd^6 \quad (1 - 17)$$

Substituting this value for B into the expression for U_{total} gives the value of the total potential energy at the minimum or

$$U(d) = -\pi NC/(9d^3) \quad (1 - 18)$$

For this value of the potential energy, the equilibrium distance, d, is 0.765 r where r is the radius of the adatom. The weak assumptions of this approach are

1. The finite summations of the potentials over the j pairs are replaced by the integrals.
2. The repulsion potential energy expressions are not known with complete certainty.

This theory applies to nonpolar molecules on nonmetallic, nonpolar solids. Extensions of this theory to metallic adsorbents (but nonpolar) must include the concept of electrons free to move through the bulk. Since these electrons are not fixed to atoms located at lattice points, the mechanism of dipole formation presented previously must be reevaluated. Lennard-Jones assumes that the electron "gas" of the metal mirrors exactly the fluctuating dipole of the adatom.[4] The attractive potential energy term between a nonpolar gas atom (spherically symmetric) and its image according to Lennard-Jones is

$$U_{ij} = -\chi mc^2/(Ld^3) \quad (1 - 19)$$

where

L = Loschmidt's number = 6.06×10^{23}
d = distance of the adatom from the surface

m = mass of one electron

c = velocity of the speed of light

χ = magnetic susceptibility of the gas

This expression for the attractive forces when combined with the previously mentioned expression for the repulsive forces is used to describe the total potential for interactions with metallic nonpolar adsorbents.

Example 1   Calculate the heat of physisorption for an argon atom on a copper metal crystal assuming the attractive potential is given by Eq. (1 - 19).

Solution. We may calculate only a gross estimate of the heat of sorption by this method since the repulsive forces are not included. For Ar the diamagnetic susceptibility is 18.13 x 10^{-6} /g-atom; m = 9.109 x 10^{-28} kg; c = 3 x 10^{10} cm/s; and L = 6.06 x 10^{23} /g-atom.

$$\mu = (9.109x10^{-28})(3x10^{-10})^2(18.13x10^{-6}/6.06x10^{23}) \qquad (1\text{ - }20)$$

or

$$\mu = 2.45 \times 10^{-11} \text{ ergs } = 35.4 \times 10^4 \text{ cal/mole} \qquad (1\text{ - }21)$$

$$W = 35.2 \times 10^4 \text{ cal/mole/}d^3 \qquad (1\text{ - }22)$$

Now what value of d must be used? Copper atoms pack in the metal crystal with a closest approach of 2.54 angstroms, while argon atoms pack with a closest approach of 3.84 angstroms. We assume Ar/Cu packs on the surface at the mean of these two interatomic distances or 3.2 angstroms. Using this value of d we calculate W as follows:

$$W = 10.8 \times 10^3 \text{ cal/mole} \qquad (1\text{ - }23)$$

This figure is higher than the actual heat of sorption since no repulsive forces are included. Experience shows the repulsive forces account for 40% of the attractive force; thus, the estimated heat of sorption is 6000 cal/mole. The observed energies of sorption at low temperatures are comparable to these calculations. The reader is referred to Reference 4, page 338 for tabulated values of the μ for the sorption of nonpolar molecules to the surface of metals.

The next logical extension of this theory involves the sorption of a nonpolar molecule to the surface of a polar crystal. Here, the polarity of the crystal substrate will induce a dipole to the nonpolar adsorbing gas molecules. The induced dipole moment, δ, of the molecule is

$$\delta = \alpha F \qquad (1 - 24)$$

where

α = polarisability of the adatom
F = electric field intensity

If F is constant throughout the volume of the adsorbing molecules, the potential energy of the induced dipole is

$$U_i = -\int \delta \, dF = -\int \alpha F \, dF = -0.5\alpha F^2 \qquad (1 - 25)$$

One is tempted to include this inductive potential energy together with the potentials of attraction and repulsion, assuming these forces to be independent. However, some reflection would indicate the very large ionic charges localized at the lattice points of the crystal could very well perturb the symmetry of the adatom electron orbitals. Such perturbations in the electron densities of the adsorbing molecule will most probably disturb the creation of the weak, instantaneous dipoles. Another problem is the uncertainty of the adatom position relative to the surface ions; the position of the adatom (directly above an ion, or between two ions in the "valley") bears directly on the value for the U_i and the dispersion potential. In the valley position between two ions, $U_{dispersion}$ is minimized, but the potentials for the inductive forces are favored when the adatom is centered above the smaller one of the lattice ions. Clearly, this problem of sorption to ionic crystals cannot by solved in general, and special cases must be developed for the solid of interest.

Lennard-Jones shows calculations (Fig. 1 - 2) of potential curves of noble gases in contact with polar crystals. For each system, two calculations are made: (a) noble gas atoms brought towards crystal above a lattice point (curves 1 and 3), and (b) noble gas atoms brought towards crystal above the midpoint of the lattice points (curves 2 and 4). For each calculation the estimate of the heat of sorption is taken to be the minimum value of the potential energy curves. The estimates are as follows [used with permission from Royal Society of Chemistry, Reference 4]:

Atom/Substrate	Heat of Sorption (cal/gm atom)	r_a, angstroms
Ar/KCl lattice pt.	-1100	3.61
Ar/KCl midpoint	-1600	3.08
Ne/NaF lattice pt.	- 400	2.90
Ne/NaF midpoint	- 500	2.50

These potential energy curves show the effect of the surface configuration upon the depth of the surface potential and the equilibrium distance from the surface.

Moreover, the type of crystal and the type of adatom has a significant effect upon the potential energy function. The midpoint configuration for the adatom assures a closer interaction with the surface and a larger heat of sorption.

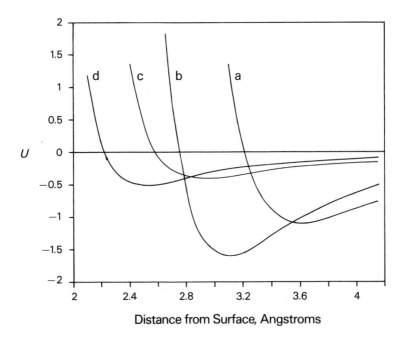

Distance from Surface, Angstroms

Figure 1 - 2 Potential Energy of Noble Gases above Polar Crystals

    a. Ar/KCl above Lattice Points
    b. Ar/KCl above Midpoints of Lattice
    c. Ne/NaF above Lattice Points
    d. Ne/NaF above Midpoints of Lattice
Used with permission from J. E. Lennard-Jones, *Trans. Farad. Soc.*, **28**, (1932), p. 340.

Polar gases can be treated after a fashion if certain assumptions are made. Consider a gas having a permanent dipole, say a, which is placed in a constant electric field of intensity, F. The potential energy of the molecule is

$$U = -aF \cos\theta \qquad (1 - 26)$$

where θ is the angle subtended by the directions of the dipole and the electric field vectors. Calculations for the gases SO_2, NH_3, CO_2, and NO_2 on barium fluoride show the dispersion and electrostatic forces make equal contributions to the total potential energy.[5] For strictly non-ionic substrates (hence, $F = 0$) the electrostatic contribution to the potential is naturally zero. These polar gases can induce small dipoles in formerly nonpolar solids. Metals give rise to an induced dipole (image dipole) worthy of consideration. Here, we assume the metal is completely polarized[6] forming the image dipole of potential:

$$U_i = -a^2/[8d^3(1 + cos\theta)] \qquad (1 - 27)$$

where θ is the angle subtended by the dipole vector and the surface normal, and d is the distance from the center of the dipole to the surface.

Once the potential energy function is known in terms of the equilibrium position from the surface, d, the heat of sorption may be calculated as the difference between the potential energies at d and infinity.

Quantum Mechanical Treatment of Sorption

The quantum mechanical description of the potential energy function between the adsorbate/adsorbent assumes that a chemical bond is formed at a discrete location by overlap of molecular orbitals in the adsorbate/adsorbent complex. The potential energy of the complex is recalculated for different values of the intermolecular distance, r, between the adsorbate/adsorbent. The heat of adsorption is the difference in the energy at infinite separation and at the intermolecular distance minus the zero point energy of vibration. In the following we review the basic concepts of quantum mechanics and describe several empirical and ab initio models.

The quantum mechanical problem involves the solution of the Schrodinger wave equation given the Hamiltonian for the system of interest. The following equation is the time dependent form; whereas most practical problems involve the time independent equation (i.e., $dY'/dt = 0$):

$$(2\pi i)H'Y' = -h(dY'/dt) \qquad (1 - 28)$$

where

H' = Hamiltonian operator
Y' = wave function, eigen vector
i = $(-1)^{0.5}$
h = Planck's constant

The Hamiltonian, H', is a function of 3N nuclear spatial coordinates (N); 3n electron spatial coordinates (n); 3N nuclear momentum coordinates (Q); 3n electron momentum coordinates (q); and time (t) for the N nuclei and n electrons. $Y'(N,Q,n,q,t)$ is the complete time dependent wave function which may be rewritten as the product of a stationary wave function, $Y(N,Q,n,q)$, and a transient wave function, $\psi(t)$.

$$Y'(N,Q,n,q,t) = Y(N,Q,n,q)\ \psi(t) \tag{1 - 29}$$

If this expression of Y' is substituted into Eq. (1 - 28) we have

$$H'(N,Q,n,q,t)\ \{Y'(N,Q,n,q)\ \psi(t)\} = Y(N,Q,n,q)\ (-h/2\pi i)\ d\psi(t)/dt \tag{1 - 30}$$

Now rewriting the H' operator as the sum of a stationary and transient operator gives

$$H'\{Y,\psi\} = H\hat{}(N,Q,n,q)\ \{Y\ \psi\} + H'(t)\{Y\ \psi\} = \psi H\hat{}Y + Y\ H'\psi \tag{1 - 31}$$

If we equate stationary and transient parts the following two equations are the result:

$$\psi(t)\ H\hat{}(N,Q,n,q)\ Y(N,Q,n,q) = \psi(t)\ E\ Y(N,Q,n,q) \tag{1 - 32}$$

$$Y(N,Q,n,q)\ E\ \psi(t) = Y(N,Q,n,q)\ (-h/2\pi i)\ d\psi(t)/dt \tag{1 - 33}$$

Equation (1 - 32) is the time independent Schrodinger equation; whereas Eq. (1 - 33) is the time dependent Schrodinger equations having the following solution:

$$\psi(t) = \exp(-2\pi i E_t/h) \tag{1 - 34}$$

These equations specify the total energy of the nuclei, E_n, and the electrons, E_{el}. Often it is of interest to know the electronic energy relative to that of the nuclei. This may be represented mathematically as the following:

$$H\hat{}(N,Q,n,q) = H(N,n,q) + H''(Q) \tag{1 - 35}$$

It may be assumed that the wave function may be separated as follows:

$$Y(N,Q,n,q) = Y_{el}(N,n,q)\ Y_n(N,Q) \tag{1 - 36}$$

Once again the separation of variables is sought by substituting Eq. (1 - 36) into Eq. (1 - 28) to give

$$H\hat{}(N,Q,n,q)\ Y(N,Q,n,q) = H(N,n,q)\ (Y_{el}(N,n,q)\ Y_n(N,Q)) +$$

$$H''(Q)\ \{Y_{el}(N,n,q)\ Y_n(N,Q)\} \tag{1 - 37}$$

Now using the definitions of the individual parts leads to

$$Y_n(N,Q) \, H(N,n,q) \, \{Y_{el}(N,n,q)\} = Y_n(N,Q) \, E_{el} \, Y_{el}(N,n,q) \tag{1 - 38}$$

or

$$H(N,n,q) \, Y_{el}(N,n,q) = E_{el} \, Y_{el}(N,n,q) \tag{1 - 39}$$

and

$$H''(Q) \, \{Y_{el}(N,n,q) \, Y_n(N,Q)\} = Y_{el}(N,n,q) \, E_n \, Y_n(N,Q) \tag{1 - 40}$$

But

$$E = E_n + E_{el} \text{ and } E_n = E - E_{el} \tag{1 - 41}$$

so that

$$H''(Q) \, \{Y_{el}(N,n,q) \, Y_n(N,Q)\} = Y_{el}(E - E_{el}) \, Y_n \tag{1 - 42}$$

which may be rearranged as the following:

$$[H''(Q) + E_{el}] \, Y_n(N,Q) = E \, Y_n(N,Q) \tag{1 - 43}$$

Equation (1 - 39) is referred to as the electronic Schrodinger equation which depends upon 3n electron coordinates and 3N nuclear coordinates (parametrically). The process of separating the nuclear and electron wave functions is called the Born-Oppenheimer approximation.[7a, 7b, 7c, 7d, 7e]

For the case of an adsorbing solid, the Hamiltonian has many terms involving the sum of kinetic and potential energies $(T_{ij} + V_{ij})$ of all the electrons and nuclei of all the atoms in the system:

$$H = \sum (T_{ij} + V_{ij}) \tag{1 - 44}$$

Of course the corresponding solutions to the wave function show many eigen vectors (E_i) and eigen functions (Y_i). The potential energy function is just the sum of the energies over all the particles contained in the system. A deeper appreciation of this problem is gained from a consideration of the individual terms comprising T_{ij} and V_{ij}. The kinetic energy terms involve the motion of the electrons and nuclei of each atom. Fortunately those terms describing the nuclei are numerically inferior to the corresponding terms for the electrons and thus they are neglected. The potential energy terms, V_{ij}, describe the forces of repulsion between electrons and between nuclei as well as the forces of attraction between an electron and the nuclei. The nuclear repulsion terms are neglected in some models next the electron/electron repulsion and the attractive forces. Thus, neglecting all nuclei repulsion terms we have

$$T_{ij} + V_{ij} = (h/2)m_{ij}\nabla^2 + \Sigma(e^2/r_{ik}) - e^2Z/r_{ij} \qquad (1 - 45)$$

The first term is the kinetic energy operator for the *ith* electron in the *jth* atom; whereas the summation expresses the electron/electron repulsion terms between the *ith* electron in the *jth* atom and all other electrons in the system. The third term is the *ith* electron attraction for the *jth* nuclei having atomic number Z. If the electron/electron repulsion terms are omitted from the Hamiltonian, then solutions are readily available.

The omission of the electron/electron repulsion terms in the Hamiltonian follows from the independent particle theory.[8] This model assumes that one particle in a system of many experiences no forces from the others. As such, the Hamiltonian may be written for the *N* particles as

$$H(1,2,...,N) = \Sigma H(i) \qquad (1 - 46)$$

where $H(i)$ is the Hamiltonian for the ith particle. The wave function is the product of *N* single-particle, molecular orbital wave functions.

$$Y(1,2,...,N) = \psi(1)\psi(2)...\psi(N) \qquad (1 - 47)$$

Now the Schrodinger equation becomes

$$H(1,2,...,N) \; Y(1,2,...,N) = E \; Y(1,2,...,N) \qquad (1 - 48)$$

$$(H(1) + ... + H(N))(\psi(1) ... \psi(N)) = E(\psi(1) ... \psi(N)) \qquad (1 - 49)$$

Equation (1 - 49) may be decomposed into *N* single-particle equations of the type

$$H(i) \; \psi(i) = e(i) \; \psi(i) \qquad i = 1,2,...,N \qquad (1 - 50)$$

where $e(i)$ is the single-particle energy corresponding to the single-particle wave function, $\psi(i)$. In the case of the electron energy relative to the nuclear energy, $e(i)$ is referred to as the one-electron energy for the *ith* moleculear orbital. The total energy of the system is merely the sum of the one-electron energies of the states occupied by electrons times the occupation number, $n(i)$.

$$E = \Sigma n(i) \; e(i) \qquad (1 - 51)$$

The extended Huckel molecular orbital theory (EHMO) is an independent particle model which includes only the valence electrons of the atoms/ions in the calculation. As before, the total *N*-electron wave function is the extended product over *N* single electron molecular ortibal wave functions. Each molecular orbital is

Chapter 1 Picture of the Fluid-Solid Interface

fabricated by the LCAO procedure from a basis set of N' atomic orbitals, A_i, about the M atoms/ions.

$$\psi(a) = \Sigma C_{ia} A(i), \quad \text{for } a = 1,...,M,...,N' \qquad (1 - 52)$$

As with many variational calculus methods, one variable, in this case the total energy, is minimized with respect to the slack variable, here C_{ia}.

$$(\partial E/\partial C_{ia}) = 0, \quad \text{for } i = 1,...,N' \text{ and } a = 1,...,M \qquad (1 - 53)$$

For the present case, minimization of E leads to a minimization of the individual orbital energies to produce a set of equations represented in matrix notation as

$$[H_{ij} - S_{ij} E] = 0, \quad \text{for } a = 1,2,...,N',$$

$$i = 1, 2,..., N' \text{ and } j = 1, 2,..., N', \qquad (1 - 54)$$

where H_{ij} is the resonance integral of the wave functions Y_i, Y_j over the Hamiltonian, H and S_{ij} is the overlap integral of the wave functions Y_i, Y_j. Equation (1 - 54) has a nontrivial solution when the $N' \times N'$ determinant of the secular matrix is zero.

$$[H_{ij} - S_{ij} E] = 0 \qquad (1 - 55)$$

The roots of the resulting N' order algebraic equation are the orbital energies, $e(i)$. The $(N' - 1)$ coefficients, C_{ia}, are uniquely determined when the values of $e(i)$ are substituted into the secular equations generated from the secular matrix. Normalization of the wave functions results in the unique assignment of values for C_{ia}. The MO's are filled two electrons at a time beginning with the orbital of lowest energy. Degenerate orbitals are filled according to Hund's Rule of Maximum Multiplicity only if these are the highest occupied molecular orbitals (HOMO), otherwise Hund's Rule is not observed.

The resonance integrals, H_{ij}, are calculated from the overlap integral, S_{ij}, and the coulomb integrals, H_{ii} and H_{jj}. The coulomb integrals are specified from data of valence state ionization potentials or the equivalent spectroscopic data.[5] One of three methods may be used to calculate H_{ij}:

1. $H_{ij} = -KS_{ij}$
2. $H_{ij} = -KS_{ij} (H_{ii} + H_{jj})/2$
3. $H_{ij} = -KS_{ij}(H_{ii} H_{jj})^{0.5}$

The Huckel constant, K, is found in practice to vary between 1.7 and 2.0. The overlap integral is calculated directly from the data of spatial coordinates and the choice of the basis set atomic orbitals. At present two sets of basis functions are used.

1. Slater type atomic orbitals with effective nuclear charges[9, 10]
2. Analytical atomic self-consistent field (SCF) orbitals

Thus, the input data for an EHMO calculation are as follows:

1. The atom/ion coordinates in the molecule or catalyst
2. The number of valence electrons for each atom/ion
3. Atomic orbital exponents for specifying the basis set AO's
4. The valence state ionization potential for each AO
5. The Huckel parameter
6. The charge on the ion/catalyst.

In the next paragraphs we describe some of the common quantum mechanical orbital calculations. The exact treatments are used for simple molecules; whereas the semiempirical calculations are often used for large molecules and surface energy calculations. These techniques employ the same energy minimization used in the EHMO; thus, we will not repeat the details given earlier. Instead we focus the discussion on the assumptions and simplifications inherent to each calculation.

The CNDO (complete neglect of differential overlap) represents electron/electron repulsion terms in the Hamiltonian by empirical parameters. Moreover, only the valence electrons are included in the self-consistent field process. The core electrons are included as an empirical parameter in the monoelectric integrals (H_{ii} of the EHMO technique). Finally, acentric orbitals about each nucleus are approximated by s-orbitals. This model is more accurate than the EHMO; however, the CNDO model requires more time to solve the problem than the EHMO procedure.

The SCF model represents all repulsion terms without any empirical parameters and the integrals are calculated for each iteration of the minimization. For an n-particle problem, the number of integrals that must be calculated is on the order of n^4! Clearly, the SCF routine can only be employed for simple molecules.

Example 2

We may compare the quantum mechanical predictions by several models by calculating the intermolecular potential of simple molecules. Consider first the calculations for molecular hydrogen by the CNDO and EHMO techniques. Hydrogen is chosen to demonstrate the simplest molecular two-electron problem. The CNDO method should predict the molecular properties better than the EHMO method since the Huckel method does not model electron/electron repulsion and the CNDO method does. Figure 1 - 3 shows the binding energy of hydrogen versus the internuclear distance for two models: CNDO and EHMO.

Whereas the CNDO calculation predicts an optimum distance between the two nuclei of 0.75 angstroms, the EHMO calculation does not show any optimum distance. This result for the EHMO is expected since only attractive terms in the Hamiltonian have been included in the calculation. Moreover the binding energies predicted by the EHMO method are much greater (e.g., about 35 eV) than what the CNDO model predicts (about 5 eV). Other models for the hydrogen molecule are reported in the literature and we show here the equilibrium distances and binding energies: Heitler-London (0.869 angstroms/3.140 eV); MO (0.85 angstroms/2.681 eV). The experimental values for internuclear distance/energy is 0.74 angstroms, 4.75 eV. Thus, the CNDO calculation predicts the equilibrium distance well but overestimates the binding energy.

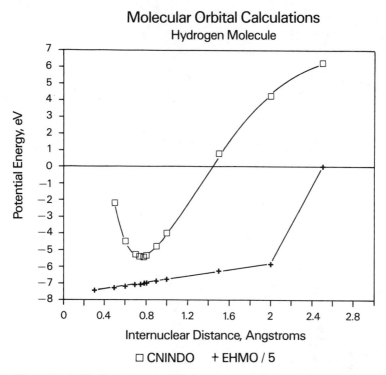

Molecular Orbital Calculations
Hydrogen Molecule

Internuclear Distance, Angstroms

□ CNINDO   + EHMO / 5

Figure 1 - 3 Binding Energy of Hydrogen versus Internuclear Distance

Example 3

The CNDO, EHMO, and SCF models were compared for the prediction of the intermolecular potential of the protonated cyclopropane system (see Fig. 1 - 4). The SCF results are reprinted here with permission from the American Chemical Society: Petke, J. D. and J. L. Whitten, *J. Am. Chem. Soc.*, **90:13**, (1968), p. 3338. The trajectory of the proton in this system is a path along the negative y-axis. Figure 1 - 5 gives the predictions of the three models as a function of proton position. The abscissa values in Fig. 1 - 5 are the y-coordinates of the proton; whereas the ordinate values are the binding energies for the proton. These proton binding energies were calculated by subtracting the total energy of the system at intermediate y coordinates from that at y = -10 angstroms.

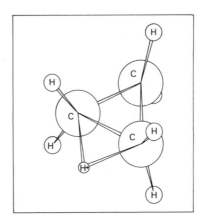

Figure 1 - 4 Schematic of Protonated Cyclopropane System

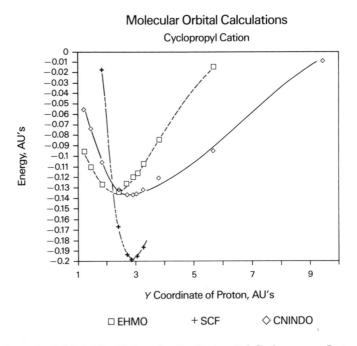

Figure 1 - 5 Model Predictions for the Protonated Cyclopropane System

All three models show extrema in the binding energies of the proton. The positions of the extrema are as follows: EHMO - 1.30, CNDO - 1.45, and SCF - 1.53 angstroms. The binding energies predicted at the extrema were as follows: EHMO, 0.13 AU's; CNDO, 1.33 AU's; SCF, -.20 AU's (1 AU = 27.2 eV). These calculations show the three models disagree on the optimum geometry for the proton in this protonated cyclopropane system. The EHMO predicts the closest approach of the H^+ to the cyclopropane ring. The EHMO model does not include any repulsion terms (electron and nuclear); thus, only the attractive part of the potential is represented. Since the proton has no electrons, then electron repulsion is less important in this calculation but nuclear repulsion is very important. The CNDO and SCF models include electron and nuclear repulsion terms. Thus, we expect the SCF and CNDO models to predict an optimum geometry for the proton which is slightly further away from the CP molecule than the EHMO model.

The EHMO model shows a minimum in this calculation as a result of the p-orbital symmetry of the carbons. As the proton approaches close to the carbons, the s orbital of the proton intrudes a space of lower p-orbital density, hence less overlap. As a result of less overlap, the system realizes less bonding and decreased stabilization. The minimum predicted by the EHMO does not represent a balance of forces that is predicted by the more exact models but is merely a consequence of less overlap.

Example 4

The sorption of ammonia to silica and silica-alumina was modeled using EHMO to define the potential energy curve for adsorption. A model for the surface was developed using cristobalite as the crystal structure for the silica and the matrix for which isomorphous substitution of Al^{+3} ion into the Si^{+4} matrix ion to model silica-alumina (see Fig. 1 - 6). The crystal was cleaved at the {001} index to create the chemisorbing surface. The sites for the base were the protons attached to the oxygens shared by the silicon and aluminum ions and the surface silanols. The potential energy curve developed from the sum of one-electron energies is plotted versus the distance of the ammonia molecule (nitrogen atom) to the proton of the sorption site. The reference potential is the sum of one-electron energies for the ammonia/crystal complex at infinite separation between the ammonia and the surface.

The following cases were developed to illustrate the weaknesses of the EHMO theory.

1. One ammonia molecule sorbed to a surface silanol of silica.
2. Two ammonia molecules sorbed to separate silanols of silica.
3. Two ammonia molecules sorbed to separate OH groups on silica alumina (Si/Al = 1).

The quantum mechanical predictions of the potential energy curve are summarized in Figure 1 - 7 as potential energy of the chemisorbed complex versus distance between the nitrogen and the surface proton. The potential energy is the difference between the one-electron energy of the complex showing a internuclear distance of d and the the one-electron energy of the complex at infinite separation. For the three cases just described the predictions describe one curve showing a minimum of -3.2 eV at an internuclear distance of 0.85 angstroms.

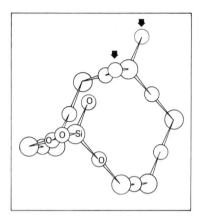

Figure 1 - 6 Model of Silica Surface

These results clearly demonstrate the failure of the EHMO theory to predict accurate potential energies since the literature of acid catalysis reveals that silica-alumina is a stronger acid than silica and for both catalysts the energy of attraction should decrease with increasing coverage of base.[11] Also, temperature programmed desorption of ammonia from an amorphous silica, Cab-O-Sil, shows an activation energy of desorption of 10 to 12 kcal/mole (0.4 to 0.5 eV) which is much lower than the energy of desorption predicted by the EHMO method.[12] The failure of the EHMO to accurately represent the activation energy for desorption of ammonia (E_d) from the silica is apparent from the assumption which neglects the electron/electron repulsion. This neglect of the repulsion terms will result in overestimating the strength of the potential (i.e., curve lies lower than expected) and the position of the equilibrium position will occur at an internuclear distance (0.85 versus 1.1 angstroms) shorter than expected. Moreover, dipole/dipole interactions between two chemisorbed molecules will not be modeled such that the variations of the heat of adsorption with coverage cannot be predicted.

The lack of electron/electron repulsion terms in the Hamiltonian is the main reason for the failure of the EHMO method to predict accurately the potential energy curve. As such, the EHMO method shares the shortcomings of the semiempirical methods such as the London theory and the others. The real value of the EHMO lies in the charge calculations. The predictions of trends in molecular properties based on localized charges can be quite good.[13]

Chapter 1 Picture of the Fluid-Solid Interface

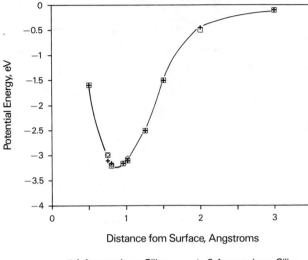

EHMO Calculation
Ammonia on Silica and Silica Alumina

□ 1 Ammonia on Silica   + 2 Ammonia on Silica
◇ 2 Ammonia on S-A

Figure 1 - 7 Potential Energy Curves for Silica System

One Ammonia Molecule on Silica
Two Ammonia Molecules on Silica; One at 0.85 Angstroms
Two Ammonia Molecules on Silica-Alumina; One at 0.85 Angstroms

Conclusions

The surface of real adsorbents is inherently more complicated than the simple models presented here; however, consideration of these models has led to greater insight to the sorption process. The dispersion theories show a common weakness in the inability to account for repulsion forces. The quantum mechanical treatment in principle can account for the repulsion forces; however, the resulting problem is intractable for all but the most simple of systems.

REFERENCES

1. Maire, G., P. Bornhardt, P. Legare, and G. Lindauer, *Proc. 7th Int. Vac. Cong. and 3rd Int. Conf. Sol. Surfaces*, Vienna, (1977), p. 86.
2a. Longuett-Higgins, H. C., *Disc. Faraday Soc.*, **40**, (1965), p. 7.
2b. Eyring, H., J. Walter, and G. W. Kimbell, *Quantum Chemistry* (New York: Academic Press, 1946).

2c. London, F., Z. Physik, **63**, (1930), p. 245.

3a. Mueller, A., Proc. Roy. Soc. (London), Ser. A, **154**, (1936), p. 624; Kirkwood, J. G., Physik. Z., **33**, (1932), p. 57.

3b. Hirschfelder, J. O., C. F. Curtis, and R. B. Bird, Molecular Theory of Gases and Liquids (New York: J. Wiley & Sons, Inc., 1954).

3c. Avgul, N. N., A. A. Isirikyan, A. V. Kiselev, L. A. Lygina, and D. P. Poshkus, Isv. Akad. Nauk., SSSR Khim. Nauk, (1957), p. 314.

3d. Hill, T. L., Adv. Catal., **4**, (1952), p. 212.

4. Lennard-Jones, J. E., Trans. Faraday Soc., **28**, (1932), p. 333.

5. Crawford, V. A. and F. C. Tompkins, Trans. Farad. Soc., **44**, (1948), p. 698.

6. Young, D. M. and A. D. Crowell, Physical Adsorption of Gases (London: Butterworth Publishers, 1962).

7a. Turner, Almon G., Methods in Molecular Orbital Theory (Englewood Cliffs, NJ: Prentice-Hall, Inc., 1974).

7b. Slater, J. C., Quantum Theory of Atomic Structures (New York: McGraw-Hill Book, Co., 1960).

7c. Daudel, R., Electronic Structures of Molecules (Oxford: Pergamon Press, 1966).

7d. Ballhausen, C. J., Introduction to Ligand Field Theory (New York: McGraw-Hill Book Co., 1962).

7e. Kauzman, W., Quantum Chemistry (New York: Academic Press, Inc., 1957).

8. Skinner, H. A. and H. O. Pritchard, Trans. Faraday Soc., **49**, (1953), p. 1254.

9. Slater, J. C., Phys. Rev., **36**, (1930), p. 57; Zener, C., ibid. (1930), p. 51.

10. Clementi, E. and D. Raidmondi, J. Chem. Phys., **38**, (1967), p. 2686.

11. Petke, J. D. and J. L. Whitten, J. Am. Chem. Soc., **90:13**, (1968), p. 3338.

12. Benesi, H. A. and B. H. C. Winquist, Adv. Catal., **27**, (1978), p. 97.

13. Beckler, R. K., Ph. D. thesis, Georgia Institute of Technology, Atlanta, GA, (1987).

14. Barrer, R. M, Zeolites and Clay Minerals as Sorbents and Molecular Sieves (New York: Academic Press, Inc., 1971), pp. 190-2.

PROBLEMS

1. The heat of adsorption per gram-mole of A (an adatom adsorbed upon a metal by ionic chemisorption) is given by

$$Q_o = [-I + \Phi + (e^2/4\pi\epsilon_0 z^*)]N_a \qquad (1 - 56)$$

where

$$
\begin{aligned}
I &= \text{ionization potential} \\
\Phi &= \text{work function of the metal} \\
e^2/4\pi\epsilon_0 z^* &= \text{energy of attraction of the adatom to the metal} \\
e &= \text{charge on the electron}
\end{aligned}
$$

$$\epsilon_0 = \text{permittivity of a vacuum}$$
$$N_a = \text{Avogadro's Number}$$

Consider the ionosorption of sodium vapor upon the surface of tungsten as given by the following:

$$W + Na \rightarrow W\text{-}Na^+ \qquad (1 - 57)$$

Calculate Q_0 for the case where $N_a\Phi = 104$ kcal/mole and $z^* = 1.83$ angstroms. [Data used with permission from Trapnell, B. M. W. and D. O. Hayward, *Chemisorption* (London: Butterworth Publishers, 1964), p. 201].

2. Show that the mean interaction energy of one mole of atoms having a diameter, d, interacting with a potential energy of the form $-C/r^6$ is given by the following:

$$U = -2\pi N_a^2 C/3Vd^3 \qquad (1 - 58)$$

where

$$V = \text{volume confining the atoms}$$
$$N_a = \text{Avogadro's Number}$$

From *Physical Chemistry*, by P. W. Atkins. Copyright 1978. Reprinted with the permission W. H. Freeman & Company.

3. The expression for the second virial coefficient, B, in terms of the intermolecular potential function, $U(r)$, is given by

$$B(T) = 2\pi N_a \int_d^\infty (1 - \exp[-U(r)/kT])r^2 \, dr \qquad (1 - 59)$$

Suppose the atoms have a distance of closest approach of d and outside that range they are attracted by $-C/r^6$. You may assume that when atoms are not in contact, $U(r)/kT$ is small such that the exponential can be expanded to the first power. Find an expression for B in terms of C and d. From *Physical Chemistry*, by P. W. Atkins. Copyright 1978. Reprinted with the permission W. H. Freeman & Company.

4. Estimate the heat of adsorption for argon gas onto solid methane. Use the potential energy function, based upon the London Dispersion Theory. Give your answer with units of kcal/mole in terms of number density of adatoms/cubic angstrom. The pertinent data are as follows: $r_{Ar} = 1.92$ angstroms; $\alpha_{Ar} = 1.6$ (angstroms)^3; $I_{Ar} = 15.7$ V; $\alpha_M = 2.6$ (angstroms)^3; $I_M = 14.5$ V.

5. Using the Lennard-Jones "6-12" rule for the potential function of oxygen gas:

$$U(r) = 4\epsilon[(\sigma/r)^{12} - (\sigma/r)^6] \qquad (1 - 60)$$

determine the radius at $300^\circ K$ and 1 atm which minimizes this function.

6. The interaction potential, based on the dispersion theory, for a pair of molecules showing interactions of higher order than dipole/dipole is given by the Kirkwood-Mueller approximation as

$$U_t = \zeta_d + \zeta_r \qquad (1 - 61)$$

where

$$\zeta_d = -A_{ij}(6)/r_{ij}^6 - A_{ij}(8)/r_{ij}^8 - A_{ij}(10)/r_{ij}^{10} \qquad (1 - 62)$$

and

$$\zeta_r = B_{ij}/r_{ij}^{12} \qquad (1 - 63)$$

For the interaction between methylene groups of hydrocarbons and substituted sodium atoms in a zeolite A lattice, the following constants[14] are given:

$r(CH_2) = 2.00$ angstroms and $r(Na) = 0.95$ angstroms
$A_{ij}(6)$ = 2.04×10^{-46} kcal-cm^6/mole
$A_{ij}(8)$ = 2.6×10^{-62} kcal-cm^8/mole
$A_{ij}(10)$ = 4.0×10^{-78} kcal-cm^{10}/mole
B_{ij} = 6.61×10^{-91} kcal-cm^{12}/mole

Calculate the interaction potential, U_t, for these conditions. Used with permission from R. A. Barrer, *Zeolites and Clay Minerals as Sorbents and Molecular Sieves* (New York: Academic Press, Inc, 1971), pp. 190-2.

CHAPTER 2

MACROSCOPIC DESCRIPTION

OF SORBATE/SORBENT INTERACTIONS

The discussion of sorption physics at fluid/solid interfaces described the sorption phenomena on the molecular scale in terms of the potential energy function. Heats of sorption were calculated from a knowledge of this potential energy function at equilibrium distance, d, and at infinite separation. The main weakness of these theories is the incomplete description of the potential function (e.g., repulsion term in the dispersion forces). In the macroscopic description of sorbate/sorbent interactions, we ignore the molecular fine structure of the system and invoke the approach of equilibrium thermodynamics to calculate the heat of adsorption.

In the macroscopic approach of equilibrium thermodynamics, our ignorance of the proper potential functions to be used (resulting from an uncertainty of the surface microstructure) is replaced by our uncertainty regarding the use of the property to describe the heat of adsorption. Clearly, the derivative of the heat flow with respect to the number of moles adsorbed will be involved; however, the system restraints (x) are not obvious:

$$(dQ/dn)_x$$

For example, the constraint variables could be T, V, surface coverage, and so on. A brief review of the first law applied to adsorbing systems now follows.

In the case of isothermal heats of adsorption, the heat liberated at constant volume has been called the differential heat of adsorption, q. Usually, the heat flow was measured in a calorimeter with the conditions of constant gas volume (V), constant adsorbed phase volume, and constant adsorbent surface area (A). Invoking the first law of thermodynamics to this closed systems yields

$$q = \{d[E_g - E_s]/dn\}_{Vs,T,A} = -\Delta E \qquad (2 - 1)$$

The delta operator has the sense of solid phase energy, E_s, minus the gas phase energy, E_g (i.e., final minus initial states). If the adsorption occurs at constant pressure (i.e., variable volume in a closed system) the first law gives the differential heat of adsorption (at constant temperature and surface area) as

$$q = \{d[H_g - H_s]/dn\}_{p,T,A} = -\Delta H \qquad (2 \text{ - } 2)$$

These relationships are just the surface analogs to the familiar ones derived from the first law for closed systems involving volume-related variables. A less familiar derivation of the isothermal heat of adsorption involves a constant pressure process which produces both volume and surface work terms. As such it is desirable to define the total energy of the system as

$$E'_s = E_s + \Phi A \qquad (2 \text{ - } 3)$$

Now the appropriate differential heat of adsorption is given by

$$q_e = \{d[E_g - E'_s]/dn\}_{T,V,\Phi} = -\Delta E' \qquad (2 \text{ - } 4)$$

This unfamiliar differential will be treated according to the method outlined by Gibbs[1a] in the next section beginning with a definition of Gibbs Adsorption.

Gibbs Adsorption[1b, 1c, 1d]

Consider now a one-component gas phase of volume, V_g, and a solid phase of volume, V_s, which are in equilibrium at temperature T and pressure p. We may define the total internal energy, E, of this system as

$$dE = TdS - pdV + \mu_s dn_s + \mu_g dn_g \qquad (2 \text{ - } 5)$$

where

n^o_g = total number of moles of gas initially = $V_g c^o_g$
n_s = total number of moles of solid
V_g = total void volume
c_g = molar concentration of bulk gas following adsorption
a = surface area of solid
Γ = Gibbs adsorption = moles adsorbed/surface area
Γ = $(n^o_g - V_g c_g)/a$
S = entropy of real system
S' = entropy of solid in the sorbate-free state
s_g = entropy of bulk gas/unit volume
S'' = surface entropy = $(S - S' - V_g s_g)/a$
Φ = spreading pressure = $-n^o_g(\mu_s - \mu'_s)/a$
μ_s = average chemical potential of adsorbing solid
μ'_s = average chemical potential of adsorbate-free solid

We write the total internal energy of the system as the integrated form

$$E = TS - pV + \mu_s n_s + \mu_g n_g \qquad (2 \text{ - } 6)$$

From this relationship plus the total derivative dE,

$$dE = T\,dS - p\,dV + S\,dT - V\,dp + \mu_s dn_s + n_s\,d\mu_s + \mu_g\,dn_g + n_g\,d\mu_g \qquad (2 \text{ - } 7)$$

and the differential expression, Eq. (2 - 5), the Gibbs-Duhem Equation for the adsorption system may be derived:

$$0 = S\,dT - V\,dp + n_s\,d(\mu_s) + n_g\,d(\mu_g) \qquad (2 \text{ - } 8)$$

where

S = the entropy of the *total* system

V = system volume of the *total* system

$n_g = V_g c_g$

Reexpresssing Eq. (2 - 8) as two separate equations dealing with the gas phase *only* and the solid phase *only* gives

$$solid: \quad S'\,dT - V_s\,dp + n_s\,d(\mu') = 0 \qquad (2 \text{ - } 9)$$

$$gas: \quad V_g s_g\,dT - V_g\,dp + c_g V_g\,d(\mu_g) = 0 \qquad (2 \text{ - } 10)$$

Gibbs asserts the properties of the adsorbed phase are merely those expressed by the difference of total properties minus the sum of the gas and the *sorbate-free* solid. This procedure suggests the sum of Eqs. (2 - 9) and (2 - 10) should be subtracted from Eq. (2 - 8).

$$(S - S' - V_g s_g)\,dT - (V - V_g - V_s)\,dp + n_s(d\mu_s - d\mu') + (n_g - c_g V_g)d(\mu_g) = 0 \qquad (2 \text{ - } 11)$$

but

$$V - V_g - V_s = 0 \qquad (2 \text{ - } 12)$$

This equation may be recast into a more easily recognized form if we let

$$aS'' = (S - S' - s_g V_g) \qquad (2 \text{ - } 13)$$

$$-a\,d\Phi = (n_s\,d[\mu_s - \mu']) \qquad (2 \text{ - } 14)$$

$$a\Gamma = (n^o_g - c_g V_g) \qquad (2 \text{ - } 15)$$

which results in

$$aS'' \, dT - a \, d\Phi + a\Gamma \, d(\mu_g) = 0 \qquad (2 \text{ - } 16)$$

Equation (2 - 16) is the surface analog to the more familiar Gibbs-Duhem equation for homogeneous phases

$$S \, dT - V \, dp + \Sigma n_i \, d\mu_i = 0 \qquad (2 \text{ - } 17)$$

For the surface/sorbate Gibbs-Duhem equation, the spreading pressure Φ plays the same role as total pressure p and the sorbate loading per unit surface, Γ, is like n_i. The case at hand has only one component which may be partitioned between the phases. A useful variation of Eq. (2 - 16) is obtained by dividing through by the sorbate loading to give

$$S'' \, dT - a' \, d\Phi + d(\mu_g) = 0 \qquad (2 \text{ - } 18)$$

Here

S'' = surface entropy/mole sorbate

a' = surface area/mole of sorbate = $1/\Gamma$

The spreading pressure may be easily related to changes in the gas phase fugacity at isothermal conditions

$$d\Phi = \Gamma \, du_g = \Gamma RT \, d(\ln f) \qquad (2 \text{ - } 19)$$

Integration between the fugacity limits of 0 to f yields

$$\int d\Phi = RT \int_0^f \Gamma \, d(\ln f) \qquad (2 \text{ - } 20)$$

At the lower limit of zero fugacity the sorbate loading will be identically zero; thus we have

$$\Phi = RT \int_0^f \Gamma \, d(\ln f) \qquad (2 \text{ - } 21)$$

Γ is extracted from the experimental data of sorbent loading per mass of adsorbent, $c = \Gamma/MA$, the molecular weight of the sorbate, M, and the specific surface area of the adsorbent, A. For simplicity, a new variable is defined $\phi = \Phi/MA$ such that

$$\phi = RT \int_0^f c \, d(\ln f) \qquad (2 \text{ - } 22)$$

where ϕ is the spreading pressure per gram of adsorbent. It is often useful to devise a reversible process to mimic a real process such that calculations can be performed on the reversible process. This calculation on the reversible process represents the properties of the real system if it were to proceed at a very slow rate. The following will illustrate the connection between ϕ and the enthalpy of adsorption.

Consider the idealized sorption process to model the isothermal sorption of a pure gas at a constant pressure p. For this idealized process, a two-chambered device is used to maintain the gas at T and p by a frictionless piston as it is metered through a valve into the adsorption cavity containing the adsorbent (Fig. 2 - 1). In the adsorbing cavity the pressure, p^*, is only slightly lower than p, and exchange of energy to the enviroment allows the temperature to be held constant at T. At that time when $p^* = p$ the sorption process is terminated. It is assumed the quantity of gas originally contained in the piston is more than that which is to be sorbed.

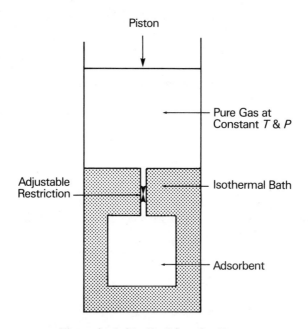

Figure 2 - 1 Idealized Sorption Process

Reprinted with permission from C. F. Snyder and K. C. Chao, *Ind. and Eng. Chem. Fund.*, **9**, p. 439. Copyright 1970, American Chemical Society.

Now we write the first law for the entire system for the adsorption of one mole of gas:

$$Q = E + W \qquad (2 - 23)$$

$$E = E' - E_g \qquad (2 - 24)$$

where

E' = molar internal energy of sorbate

E_g = molar internal energy of gas before sorption

This sorbate molar internal energy is determined in the same manner as the surface entropy; that is,

$$aE' = E - E_o - E_g \qquad (2 - 25)$$

Since the piston is advancing against a constant pressure, p, then the work attending the sorption of one mole of gas is

$$W = - \int p \, dV = -pV_g \qquad (2 - 26)$$

Thus the heat, Q, is

$$Q = E' - E_g - pV_g = E' - H_g \qquad (2 - 27)$$

The enthalpy of sorption is just the negative of the heat effect for an isobaric process, or

$$\Delta H_{ads} = -Q = H_g - E' \qquad (2 - 28)$$

Thus, we need only to evaluate E' in order to calculate the enthalpy of adsorption since H_g is readily determined from the p-V-T properties of the pure gas relative to a datum. As with the entropy we can write the internal energies for the adsorbate-free solid, E_o, and the gas phase, E_g, as

$$E_o = TS_o - pV_s + \mu_o n_s \qquad (2 - 29)$$

$$V_g E_g = Ts_g V_g - pV_g + \mu_g c_g V_g \qquad (2 - 30)$$

Now the internal energy of the absorbed layer, aE', is just the difference between Eq. (2 - 25) and the sum of Eqs. (2 - 29) and (2 - 30).

$$aE' = E - E_o - V_g E_g = T(S - S_o - V_g s_g) - p(V - V_s - V_g) + \mu_g (n_g - V_g c_g) + n_s(\mu - \mu') \qquad (2 - 31)$$

but

$$V - V_s - V_g = 0 \qquad (2 - 32)$$

so

Chapter 2 Thermodynamics

$$aE' = aTS' - a\phi + a\mu_g \Gamma \tag{2-33}$$

Once again dividing by $a\Gamma$ yields

$$E' = TS' - a'\phi + \mu_g \tag{2-34}$$

The objective now is to calculate the surface, internal energy from data of Gibbs adsorption at various temperatures and pressures plus the p-V-T properties of the gas phase. The starting point is Gibbs-Duhem equation for the sorbate phase, Eq. (2 - 16)

$$S' \, dT - d\Phi + \Gamma \, d\mu_g = 0 \tag{2-35}$$

for which we may calculate the surface entropy at constant spreading pressure (i.e., $d\Phi = 0$) from the temperature derivative of the *gas* phase chemical potential.

$$S' = -(\partial \mu_g / \partial T) \tag{2-36}$$

Consider now the gas phase chemical potential, μ_g, as a function of T and p.

$$d\mu_g = -s_g \, dT + \underline{V}_g \, dp \tag{2-37}$$

where the line under the gas phase volume indicates per unit mole. If we divide by dT subject to the constraint of constant spreading pressure, then

$$(\partial \mu_g / \partial T)_\phi = -s_g + V_g(\partial p / \partial T) \tag{2-38}$$

Now this expression for the temperature derivative, Eq. (2 - 38) may be put into Eq. (2 - 36) to yield the surface entropy as

$$S' = s_g - V_g(\partial p / \partial T) \tag{2-39}$$

Using the compressibility factor notation for V_g we have

$$S' = s_g - zRT(\partial p / \partial T)/p \tag{2-40}$$

The surface internal energy can be expressed as

$$E' = Ts_g - zRT(\partial lnp / \partial T) - a'\Phi + \mu_g \tag{2-41}$$

Recalling the expression for the enthalpy of sorption as $\Delta H_{ads} = H_g - E'$,

$$\Delta H_{ads} = H_g - Ts_g + zRT^2(\partial lnp / \partial T) + a'\Phi - \mu_g \tag{2-42}$$

but the chemical potential for the gas phase is

$$\mu_g = H_g - Ts_g \qquad (2 - 43)$$

so that

$$\Delta H_{ads} = zRT^2(\partial lnp/\partial T) + a'\Phi \qquad (2 - 44)$$

Example Problem   Sorption of a Pure Gas.   The evaluation of the terms in Eq. (2 - 44) needs clarification. The following example illustrates how data of adsorbent loading, in units of moles of gas per gram of sorbent, as a function of gas temperature and fugacity may be used to calculate the enthalpy of adsorption. The procedure for this calculation for the adsorption of a pure gas is as follows.

1. At a specific condition of temperature and pressure, determine the sorbate loading from data.

2. Determine the spreading pressure as a function of pressure, p, from Eq. (2 - 21). Here the sorbent loading (Γ) must be curve-fit to the gas phase fugacity f. The gas phase fugacity is determined from the gas phase equation of state, z. Plot this spreading pressure versus p.

3. Determine the derivative $(\partial lnp/\partial T)$ at constant spreading pressure by numerical differentition of the spreading pressure versus p data at various temperatures, T. This derivative is obtained by choosing values of the spreading pressure for which the variation in p versus T may be obtained (see Fig. 2 - 2). Values of the derivative are determined for several values of the spreading pressure.

4. The enthalpy of adsorption as a function of sorbent loading, ΔH_{ads}, is obtained from the following information:
   a. Sorbent loading Γ, hence c, and thus ϕ.
   b. Equation of state for the gas, z, hence ϕ versus $f(p,T)$.
   c. Variation of p versus T at constant spreading pressure.

For a numerical example we take that reported by Snyder and Chao.[2] The statement of the problem now follows.

Calculate the heat of adsorption of methane at $200°F$ and 5350 mm Hg. From the data of Ray and Box,[3] for the sorption of methane on activated charcoal, Snyder and Chao determined the value of c (sorbent loading) to be 41.7 cc(NPT)/gram. Moreover, they correlated these data of loading versus fugacity as a power series-logarithmic function where

$$c = A_0 + A_1 f + A_2 f^2 + A_3 f^3 + A_4 f^{0.5} + A_5 ln(f) \qquad (2 - 45)$$

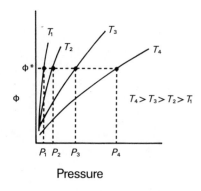

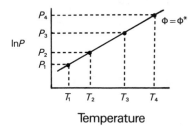

Figure 2 - 2 Spreading Pressure Curves

For small values of the fugacity they used Henry's Law as

$$c = f/H \qquad\qquad (2 \text{ - } 46)$$

where H is the Henry's Law constant. These correlations were integrated by Eq. (2 - 22) to evalute ϕ for the variables $f(p)$ and T. In the present case, Figure 2 - 3 shows the spreading pressure versus total pressure curve for which $\phi = 35,000$ cc-$^{\circ}$R/g at $T = 200\,^{\circ}$F. The appropriate derivative of the total pressure with temperature at $\phi = 35,000$ cc $^{\circ}$R/g is given in Figure 2 - 4 to be 7.86 x 10^{-3} $^{\circ}$R^{-1}. Upon substitution of these values of c, ϕ, and the derivative into Eq. (2 - 44) we have

$$\Delta H_{ads} = 8470 \ Btu/lb\text{-}mole \ adsorbed$$

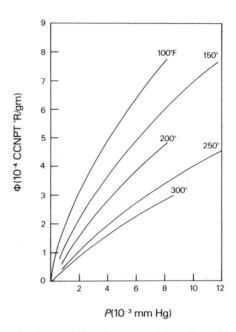

Figure 2 - 3 Spreading Pressure of Adsorbed Methane

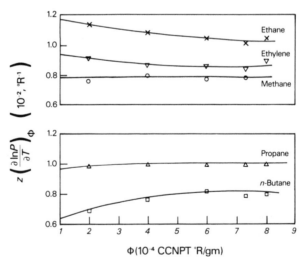

Figure 2 - 4 Derivative at Constant Spreading Pressure

Reprinted with permission from C. F. Snyder and K. C. Chao, *Ind. and Eng. Chem. Fund.*, p. 439. Copyright 1970, American Chemical Society.

Chapter 2 Thermodynamics

Conclusions

This example of finding the heat of adsorption from adsorption data is but one illustration of the equilibrium methods. In the present case the enthalpy of adsorption is calculated at constant spreading pressure; whereas the constraints may vary in other applications (e.g., the isosteric heat of adsorption is constrained to constant fractional coverage). Regardless of the equilibrium method of choice, the scientist is faced with the problem of relating the measured variable to the theoretical variable, which is usually the potential energy curve.

This discussion of the Gibbs adsorption isotherm describes a method of calculating a macroscopic parameter for example, enthalpy of adsorption from the simple measurements of adsorption capacities versus temperature and pressure. This enthalpy of adsorption is an important design parameter for packed-bed adsorption devices for the removal of carbon dioxide, water, and so on from streams. The very highly exothermic nature of these adsorptions makes it imperative to know the enthalpy of adsorption with some accuracy ($\pm$ 10%). The Gibbs approach allows for an estimate of this enthalpy through a thermodynamic "device" which correctly accounts for the volume and surface work contributions to the total system energy.

REFERENCES

1a. Gibbs, J. W., *Collected Works* (New York: Longmans Green, 1931).

1b. Hill, T. L., *Adv. Catal.*, **4**, (1952), p. 211.

1c. Young, D. M., and A. D. Crowell, *Physical Adsorption of Gases* (London: Butterworth Publishers, 1962).

1d. Chao, K. C. and R. A. Greenkorn, *Equilibrium Thermodynamics of the Fluid Phase* (New York: Marcell-Dekker, Inc., 1974).

2. Snyder, C. F. and K. C. Chao, *Ind. and Eng. Chem. Fund.*, **9**, No. 3, (1970), p. 437.

3. Ray, G. C. and E. O. Box, *Ind. Eng. Chem.*, **42**, (1950), p. 1315.

4. Kidnay, A. J. and A. L. Meyers, *Am. Inst. Chem. Eng. J.*, **12**, (1966), p. 981.

PROBLEMS

1. It is desired to study the adsorption of carbon dioxide on NaX zeolite. Table 2 - 1 gives the sorbate loading for carbon dioxide on the zeolite at two temperatures.

a. Calculate the fugacity, f, at each pressure for the two temperatures.

b. From the data of Table 2 - 1 [used with permission from Breck, D. W., *Zeolite Molecular Sieves* (New York: John Wiley & Sons, Inc., 1974), p. 611], calculate the enthalpy of sorption for 1 g-mol of carbon dioxide at 500 mm Hg and $25°C$.

		Table 2 - 1	Sorbate Loading on NaX		
$T\ (^{\circ}K)$	p (Torr)	x/m (g/g)	$T\ (^{\circ}K)$	p(Torr)	x/m(g/g)
195	0.1	0.06	298	2.0	0.06
	1.0	0.20		10.0	0.10
	6.0	0.33		100.0	0.20
	100.0	0.39		700.0	0.26

2. Gibbs defined a quantity of surface thermodynamics which is analogous to the pressure. This surface pressure is the energy to change the extent of surface area or

$$g = dW_{rev}/dA \qquad\qquad (2 - 47)$$

where dW_{rev} is the reversible work to cause an increase in the surface area of dA. Prove that

$$g = dG_s/dA \qquad\qquad (2 - 48)$$

where G_s is the surface free energy.

3. Using the results of Problem 2 - 2, derive the Kelvin equation for pressure difference at a droplet interface.

$$RT/V_m\ ln(p/p_o) = 2\gamma/r \qquad\qquad (2 - 49)$$

where

R	=	gas constant
T	=	temperature of the system, absolute
V_m	=	molar volume of the liquid at T
p, p_o	=	pressures on either side of the interface
γ	=	surface tension of the liquid at T
r	=	radius of the drop

4. Show that the volume adsorbed per unit area of sorbent is related to the pressure of the adsorbate gas by the following expression:

$$(V_{ads}/S) = (-1/RT)\ d\mu/d\ lnp \qquad\qquad (2 - 50)$$

where

V_{ads}	=	volume adsorbed
S	=	surface area of adsorbent
R	=	gas constant

T = temperature

μ = chemical potential

p = system pressure

(From *Physical Chemistry*, by P. W. Atkins. Copyright 1978. Reprinted with permission of W. H. Freeman & Company.)

5. Find the form of the adsorbate phase chemical potential in the Gibbs isotherm which yields the Langmuir isotherm. (From *Physical Chemistry*, by P. W. Atkins. Copyright 1978. Reprinted with permission of W. H. Freeman & Company.)

6. The isosteric heat of adsorption for carbon dioxide on NaX zeolite has been calculated from the Langmuir isotherm as 5.42 kcal/mole at the saturation conditions for carbon dioxide. Calculate the heat of adsorption for the conditions of Problem 1 and compare the results.

7. According to Snyder and Chao, their values of the spreading pressure, ϕ, were generally lower by 1500 cc (NPT) $^\circ$R/g compared with the results of Kidnay and Meyers.[4] If this difference is assumed to be constant, would this have any effect upon the calculated heat of adsorption? If so, what is the magnitude of the effect?

8. Derive the Gibbs adsorption isotherm.

9. Consider the sorption onto a solid for which the loading, c, at any temperature, T, is given by Henry's Law.

$$c = f/H(T) \qquad (2 - 51)$$

Assume the fugacity, f, is equal to the pressure such that Henry's Law reduces to the following:

$$c = p/H(T) \qquad (2 - 52)$$

Develop the relationship for the spreading pressure and the enthalpy of adsorption in terms of the Henry's Law constant, H, and the compressibility factor, z.

10. The Harkins-Jura Equation[1c, p. 213] is an empirical equation of state for a "condensed film" in the form

$$\Phi = b - a\sigma \qquad (2 - 53)$$

where

Φ = spreading pressure

σ = mean area occupied by an adsorbed molecule

a and b = constants

Using the Gibbs adsorption isotherm, transform the equation of state into the Harkins-Jura isotherm:

$$ln(p/p_o) = B - A/V^2_{ads} \qquad (2 \text{ - } 54)$$

CHAPTER 3

SORPTION ISOTHERMS

The two previous approaches describing sorption behavior involved (1) the potential energy between sorbate species and sorbent surface atoms, and (2) classical equilibrium thermodynamics. The goal of both approaches was to describe the heats of sorption. Here, we illustrate two more approaches to the same goal: a kinetic derivation beginning with collision theory and another thermodynamic analysis beginning with statistical mechanics. In this chapter we develop more fully the concepts of ideal and nonideal surfaces.

Sorption Isotherms from Collision Theory[1, 2, 3]

An adsorption isotherm will be derived by applying the collision theory of homogeneous gas reactions to model the interaction between gas molecules and surface atoms. We envision (Fig. 3 - 1) the adsorption/desorption equilibria between gases and solids as a reversible "reaction." This reversible reaction shows a potential energy diagram very similar to that of a chemical reaction, except the reaction "coordinate" is given by the distance between the adsorbate molecule and the surface site. Molecules having kinetic energies less than E merely rebound to the gas phase; whereas·molecules having energies greater than E and less than E' may become sorbed, thus releasing energy of adsorption equal to $-\lambda$. For the adsorption to occur there must be a collision between a gas molecule and a vacant surface site. From the collision theory of gas reactions adopted to include encounters with solid surfaces, the rate at which gas molecules strike a surface per unit area is given by z':

$$z' = 0.25n(8kT/\pi m)^{0.5} \qquad (3 - 1)$$

where

z' = collision rate = encounters/time-unit surface area
n = number density of gas molecules/cc gas = p/kT
k = Boltzmann's constant
T = Temperature, $^{\circ}K$
m = mass of a gas molecule, gm
p = partial pressure of the gas, atm

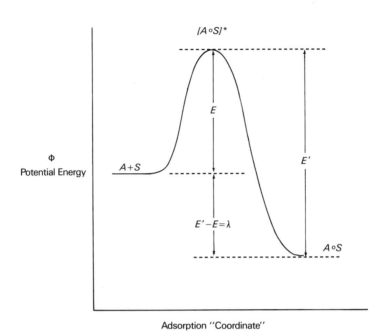

$$A+S=A \circ S$$

$[A \circ S]^*$

Φ
Potential Energy

$A+S$

E

E'

$E'-E=\lambda$

$A \circ S$

Adsorption "Coordinate"

Figure 3 - 1 Potential Energy Diagram for Reversible Sorption

This Eq. (3 - 1) represents the maximum *rate* of adsorption if every encounter with the surface leads to an adsorbed state. If the incoming molecule encounters an ideal surface with the "required" kinetic energy, then adsorption will occur. An ideal surface has each surface site showing the same energy of interaction with an adsorbate molecule. This energy is not influenced by the presence of adsorbate molecules on adjoining site. The energy required for adsorption is just the activation energy, E; molecules showing kinetic energies less than E will not be sorbed. The fraction of molecules having energies greater than or equal to E is given by $x(T,E)$:

$$x(T,E) = exp(-E/kT) \qquad (3 - 2)$$

Only those sites which are unoccupied will accept an incoming species; the fraction of sites uncovered will be denoted $f(\theta)$; whereas the fraction of sites which are covered is $g(\theta)$. Clearly, the fractions are related:

$$f(\theta) = 1 - g(\theta) \qquad (3 - 3)$$

The rate of collisions with the surface resulting in adsorption events is given by

$$z = \{collision\ frequency\}\ \{fraction\ of\ surface\ not\ covered\}$$
$$\{fraction\ of\ molecules\ having\ required\ energy\} \tag{3 - 4}$$

$$z = z'[f(\theta)]\ exp(-E/kT) \tag{3 - 5}$$

Equation (3 - 5) gives the absolute rate of adsorption, molecules/time per unit surface area based on the collision theory of gases with a plane interface. Experience shows this frequency is too high; thus we introduce a factor known as the "orientation factor." The full equation now becomes

$$z = s[p/(2\pi mkT)^{0.5}][f(\theta)]\ exp(-E/kT) \tag{3 - 6}$$

Equation (3 - 6) may be recast into a more familar form as

$$z = k_o exp(-E/kT)f(\theta)p \tag{3 - 7}$$

where

$$k_o = s/(2\pi mkT)^{0.5} \tag{3 - 8}$$

The collision theory predicts the sorption rate is linear in the adsorbate partial pressure and to show an exponential temperature dependence divided by $T^{0.5}$. The desorption rates are *postulated* to have the same form by the law of microreversibility

$$z'' = k''g(\theta)\ exp(-E'/kT) \tag{3 - 9}$$

where

k'' = desorption constant
$g(\theta)$ = a function of surface coverage
E' = activation energy for desorption

At equilibrium, the rates of adsorption and desorption are equal such that

$$s[p/(2\pi mkT)^{0.5}]f(\theta)\ exp(-E/kT) = k''g(\theta)\ exp(-E'/kT) \tag{3 - 10}$$

which upon rearrangement yields

$$g(\theta)/f(\theta) = [s/(2\pi mkT)^{0.5}](1/k'')\ exp[(E' - E)/kT]p \tag{3 - 11}$$

This function of the surface coverage, θ, is called an adsorption isotherm if one fixes the temperature, T, and varies the pressure of the adsorbate, p, to change the surface coverage. Alternatively, the pressure may be fixed and the temperature varied to change the surface covered resulting in an isobar. Equation (3 - 11) is usually simplified to yield

$$g(\theta)/f(\theta) = K_o\, exp(Q/kT)p \qquad (3 - 12)$$

where

$$K_o = s/[k''(2\pi mkT)^{0.5}]$$

Here, Q, is the negative of the heat of sorption, $-\lambda$ and it is calculated by the difference between the activation energies for adsorption and desorption E, E'. Equation (3 - 12) is a form of the Langmuir adsorption isotherm. The functions, $f(\theta)$ and $g(\theta)$, describe the stoichiometry of the adsorption event. If one sorption site is required for each sorbate molecule, then the functions are as follows:

$$f(\theta) = 1 - \theta$$
$$g(\theta) = \theta$$

Other adsorption stoichiometries are given in Table 3 - 1. Using simple functions for

Table 3 - 1 Adsorption Stoichiometries

Type of Sorption	$f(\theta)$	$g(\theta)$
One adsorbate (associative)	$(1 - \theta)$	θ
One adsorbate (dissociative)	$(1 - \theta)^2$	θ^2
Two adsorbates (competitive sorption)	$(1 - \theta_i - \theta_j)$	θ_i or θ_j
Two adsorbates (non-competitive)	$(1 - \theta_i)$ and $(1 - \theta_j)$	θ_i or θ_j

$f(\theta)$ and $g(\theta)$, one adsorbate and associative, gives the following equation where $K = K_o\, exp[Q/kT]$.

$$\theta = Kp/(1 + Kp) \qquad (3 - 13)$$

Chapter 3 Sorption Isotherms

Equation (3 - 13) is the simple Langmuir isotherm. The assumptions implicit to the present development are

1. The adsorbed species are attached to discrete sites with known stoichiometries as described by $f(\theta)$ and $g(\theta)$.
2. The activation energies E and E' and the difference $(E' - E)$ are fixed for each site and independent of coverage, θ.

These Langmuir isotherms show only limited success in modeling real adsorbents. Brunauer[4] raises two objectives to the Langmuir model as given here:

1. The apparent saturation of a surface is often observed when only a small fraction of the available surface is covered.
2. The volume of a monolayer increases with decreasing temperature.

Still other objections are noted for the heat of sorption. Consider now the simple Langmuir isotherms at temperatures T_1 and T_2 for the constant exposure pressure p. The corresponding fractional coverages, θ_1 and θ_2, are

$$\theta_1/(1 - \theta_1) = (sp/k''(2\pi mkT_1)^{0.5})\,exp(Q/kT_1) \qquad (3 - 14)$$

and

$$\theta_2/(1 - \theta_2) = (sp/k''(2\pi mkT_2)^{0.5})\,exp(Q/kT_2) \qquad (3 - 15)$$

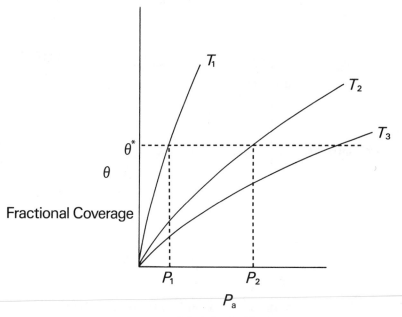

Figure 3 - 2 Isosteric Heat of Adsorption from Sorption Isotherms

These equations may be plotted θ versus p for T_1 and T_2 to give the sorption isotherms as in Figure 3 - 2. For a particular value of θ, say θ^*, the gas sorption pressures, p_1 and p_2, may be related to the temperatures, T_1 and T_2, by

$$\theta^*_{T1}/\theta^*_{T2} = [K_{o,T1}/K_{o,T2}][p_1/p_2] \, exp[(Q/k)(1/T_1 - 1/T_2)] \qquad (3 - 16)$$

Since $\theta^*(T_1) = \theta^*(T_2)$ by design and assuming the steric factors are equal at T_1 and T_2, then

$$p_1/p_2 = (T_1/T_2)^{0.5} \, exp[(Q/k)(1/T_2 - 1/T_1)] \qquad (3 - 17)$$

When T_1/T_2 is between 0.9 and 1.1 the radical is within 5% of unity; thus it can be neglected to yield

$$ln(p_1/p_2) = Q/k(1/T_2 - 1/T_1) \qquad (3 - 18)$$

The value of Q determined by Eq. (3 - 18) is called the isosteric (constant coverage) heat of adsorption. For ideal surfaces (i.e., *not* "heterogeneous") Q is independent of the coverage. For heterogeneous surfaces, Q is a function of coverage (see Fig. 3 - 3). Since the Langmuir model assumes Q is independent of θ, then the Langmuir isotherm should be restricted to model only ideal surfaces, or to heterogeneous surfaces at very low coverages.

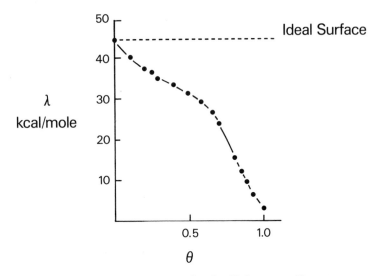

Figure 3 - 3 Heats of Adsorption for Hydrogen on Tungsten

Used with permission from B. M. W. Trapnell, *Proc. Roy. Soc.*, **A 206**, (London: Royal Society of England, 1951), p. 39.

This heterogeneity in the heats of sorption may arise from several sources:

1. Electronic redistribution in the solid as a consequence of the sorption. Here the sorbate-sorbent interaction may be attended by a formal charge transfer between the two giving rise to ionically charged sorbate molecules.[5] This process has been called ionosorption.

2. Change in the surface bonding with coverage. Spectroscopic data for CO on metals such as Pt, suggest at least two forms of chemisorbed carbon monoxide.[6, 7] The distribution of the sites is observed to change with coverage.

3. Interaction of adsorbed molecules in the formation of dipoles, quadrapoles[7] or for neutral repulsion.[6]

These effects may combine such that the heat of sorption does vary with coverage. If we allow that the adsorption enthalpies infer the energetics of the site (i.e., $Q = E' - E$) then the most energetic sites ($E' - E$ is greatest) will be "covered first" in that they will be uncovered last. The gas phase conditions dictate the frequency with which the *entire* surface will suffer collisions by adsorbate molecules; thus sites of all energies are equally probable for suffering a collision. However, not all sites show equal probability for suffering a collision that results in a site being covered against the desorption event. Increasing the *difference* in activation energy for desorption/adsorption ($E' - E$) describes a decreasing probability for desorption. Thus, high energy sites ($E' - E$ is large) will be uncovered at a rate much slower than low energy sites. As a result, the higher energy sites are covered a larger percentage of time relative to the low energy sites. Hence, the observed result that high energy sites are "covered first" in an unsteady state chemisorption experiment.

One could model a surface showing heterogeneous heats of sorption by constructing a distribution function for Q as a function of θ. The genesis of this function from fundamental considerations is not clear insofar as the desorption phenomenon is poorly understood. While the desorption event is considered to be a bond rupture process in the most general sense, it is not a classical decomposition. The "unimolecular" decompositions described by first-order processes are essentially the result of bimolecular collisions[8] with other reactant molecules or inerts to supply the energy necessary for bond activation. The second-order dependency of the reaction kinetics at very low pressures established this theory. In the context of an "immobile" surface species, the energy transfer is between the "inert" catalyst molecule and the chemisorbed reactant molecule. In lieu of a detailed mechanism, some progress has been achieved by curve-fitting of actual Q versus θ data. One particular approach is illustrated by the *Temkin* isotherm where

$$Q = -a\theta + Q_o \qquad (3 - 19)$$

is substituted into the Langmuir type isotherm and the equation solved for θ in terms of T and p. The following example develops the Temkin isotherm for associative, single site, single adsorbate sorption stoichiometry.

Example 1 Let $Q = -a\theta + Q_o$ and

$$\theta/(1 - \theta) = (K_o \exp(Q_o/kT) \exp(-a\theta/kT))p \qquad (3 \text{-} 20)$$

Taking the logarithm gives

$$ln(\theta/(1 - \theta)) = lnK_o + Q_o/kT - a\theta/kT + ln(p) \qquad (3 \text{-} 21)$$

Upon rearranging to yield an expression of θ in terms of T and p gives

$$ln(\theta/(1 - \theta)) + a\theta/kT = ln(K_o p) + Q_o/kT \qquad (3 \text{-} 22)$$

we now invoke a series expansion for $ln[\theta/(1 - \theta)]$ where $x = \theta/(1 - \theta)$:

$$ln\,x = 2[\{(x - 1)/(x + 1)\} + (1/3)\{(x - 1)/(x + 1)\}^3 + ...]\ \ for\ x > 0 \qquad (3 \text{-} 23)$$

to be reexpressed in terms of θ retaining only linear terms in the expansion

$$ln[\theta/(1 - \theta)] = 2[[(\theta/(1 - \theta)) - 1]/(\theta/(1 - \theta)) + 1] = 2[(2\theta - 1)/1] \qquad (3 \text{-} 24)$$

so that

$$a\theta/kT + 2(2\theta - 1) = ln(K_o p) + Q_o/kT \qquad (3 \text{-} 25)$$

or

$$\theta[a/kT + 4] = 2 + Q_o/kT + ln(K_o p) \qquad (3 \text{-} 26)$$

and

$$\theta = [2 + Q_o/kT + ln(K_o p)]/(a/kT + 4) \qquad (3 \text{-} 27)$$

or

$$\theta = (1/[a/kT + 4])\,ln(K_o p) + [2 + (Q_o/kT)]/[a/kT + 4] \qquad (3 \text{-} 28)$$

$$\theta = (1/a'(T))\,ln[K_o(T)p] + b(T) \qquad (3 \text{-} 29)$$

The fitted constants $a'(T)$ and $b(T)$ are characteristic of the solid/sorbate interaction (i.e., a and Q_o); whereas K_o is the Langmuir-type, adsorption equilibrium

Chapter 3 Sorption Isotherms

coefficient $[K_o = s/(k''(2\pi mkT)^{0.5})]$. The Temkin isotherm is characterized as semiempirical in that the deficiency of the Langmuir theory is "satisfied" using an empirical relationship between heat of adsorption and surface coverage. Other isotherms and the applications are listed in Table 3 - 2.

Table 3 - 2 Sorption Isotherms

Isotherms	Equation	Application
Langmuir	$\theta = Kp/(1 + Kp)$	Chemisorption and Physisorption
Freundlich	$\theta = kp^{1/n}$; (n > 1)	Chemisorption and Physisorption
Henry	$\theta = kp$	Chemisorption and Physisorption
Temkin	$\theta = (1/a') \ln(Cp)$	Chemisorption

One point must be made now regarding the validity of "repairing" the Langmuir development with a single alteration of the heat of adsorption. Data of the chemisorbed vibrational spectra of molecules such as carbon monoxide on metals corroborate the distribution of sorption energies; however, these same data clearly show different stoichiometries among the surface sites. That is, the high energy sites usually show larger site requirements per molecule (multiple site, "bridge bonding") than the low energy sites (single site). The evidence for this changing site stoichiometry is the vibrational energy of the CO bond taken together with calculations of the same assuming a model for the adsorbate molecule/site interactions.[9] The modeling of such nonideal surfaces by the collision theory requires a reexamination of the entire expression for z.

Sorptions onto Nonideal Surfaces

Consider now a quantitative description of sorption rates to a nonideal surface for which s, $f(\theta)$ and E are both functions of surface coverage. Then

$$z = p\, s(\theta) f(\theta)\, exp[-E(\theta)/kT]/\ (2\pi mkT)^{0.5} \qquad (3 - 30)$$

where $s(\theta)$, $f(\theta)$, and $E(\theta)$ will vary from point to point on the surface. In the most general sense we must develop distribution functions, Ps, Pf, and PE, of the surface elements, dS, where

$$s(\theta) = s[\theta\{Ps(S)\}] = s(S) \qquad (3 - 31)$$

and

$$f(\theta) = f[\theta\{Pf(S)\}] = f(S) \qquad (3 \text{ - } 32)$$

and

$$E(\theta) = E[\theta\{PE(S)\}] = E(S) \qquad (3 \text{ - } 33)$$

so that

$$z = p \int s(S)f(S) \, exp[-E(S)/kT]dS/(2\pi mkT)^{0.5} \qquad (3 \text{ - } 34)$$

In particular the distribution functions are not known, forcing the use of empirical relations. We now turn our attention to the function $f(\theta)$.

As given in Table 3 - 1 of this chapter, the sorption stoichiometry function $f(\theta)$ depends upon the mode of sorption (associative/dissociative) and the number of species formed upon dissociation. Let us focus on the case of finite coverage concerning a dissociative sorption process resulting in species immobile on the surface and requiring a given number of solid atoms for the process to occur. Here, $f(\theta)$ is given for the activation of a diatomic molecule by.[10]

$$f(\theta) = \{fraction \ of \ available \ sites\}\{number \ of \ available \ neighbors\}$$

$$\{fraction \ of \ neighbors \ occupied\} \qquad (3 \text{ - } 35)$$

$$f(\theta) = (1 - \theta)\{(1 - \theta)(Z)(1/[Z - \theta])\} = (1 - \theta)^2 \, Z/(Z - \theta) \qquad (3 \text{ - } 36)$$

Z is determined by the crystal structure of the surface and is equal to the number of nearest neighbors. For example, if the surface is hexagonal, cubic, closepacking then $Z = 6$ and for simple cubic packing then $Z = 4$. Two limiting cases may be identified for which simplifications are possible:

$$\theta \ less \ than \ Z: \ f(\theta) = (1 - \theta)^2$$
$$\theta \rightarrow 1: \ f(\theta) = (1 - \theta)^2 Z/(Z - 1)$$

The following example problem will serve to illustrate this idea further.

Example 2 Describe the $f(\theta)$ function for the activation of a triatomic molecule on a surface having simple cubic structure and exposing the {100} surface.

Solution A triatomic molecule will be activated into three atoms, thus requiring an ensemble of three neighboring surface atoms. For the present we assume that no preferred symmetry of the surface atoms is required. Simple cubic structure of the {100} plane will show four nearest neighbors or $Z = 4$. The probability for the first unoccupied surface atom is $(1 - \theta)$. For the second and subsequent surface sites, four are available out of $(1 - \theta)Z/(Z- \theta)$ such that the total probability of finding three unoccupied neighboring sites is the product of the three probabilities or

$$f(\theta) = (1 - \theta)(1 - \theta)(Z/(Z - \theta))(1 - \theta)(Z/(Z - \theta)) \tag{3 - 37}$$
$$f(\theta) = (1 - \theta)^3 Z^2/(Z - \theta)^2 \tag{3 - 38}$$

The extremum values are:

$$\theta \text{ less than } Z : \ f(\theta) = (1 - \theta)^3$$
$$\theta \to 1 : \ f(\theta) = (1 - \theta)^3 Z^2/(Z - 1)^2$$

Now we turn our attention to activated sorption requiring adjacent sites that results in *mobile* surface species. This mobility allows greater interspecies distances arising from the existing repulsive forces. Two observations have been made from a statistical mechanics treatment of this case:

1. The number of vacant, *adjacent* sites diminishes with the increased mobility of the sorbed species.
2. Whereas the reactivity of these mobile species towards the vacant sites is greater than for the case of immobile species.

Qualitatively, the effects tend to offset one another so that

$$f(\theta) = (1 - \theta)^3 \tag{3 - 39}$$

for the case of dissociative chemisorption of a triatomic molecule.

Kinetics of Irreversible Sorptions

Next we examine the relation of $E(\theta)$ in describing nonideal surfaces for irreversible sorptions (i.e., no desorption). Two cases may be described: (a) the *uniform* (but not ideal) surface and (b) the *nonuniform* surface.

For the uniform surface the variation of E with θ is given for all surface area elements (dS) by

$$E = E_o + a\theta \tag{3 - 40}$$

and $s(S)$ is assumed to be constant over the surface, S. The rate of sorption then is

$$z = sp(1 - \theta)/(2\pi mkT)^{0.5} \exp[-E/kT] \tag{3 - 41}$$

At sufficiently low coverages $(1 - \theta) = 1$ such that

$$z = sp/(2\pi mkT)^{0.5} \exp[-E/kT] \tag{3 - 42}$$

If the mass of the sorbing species is m, then the *rate of sorption* in terms of dq/dt (mass/time) is

$$dq/dt = (mz)N; \quad N = \text{Avogadro's Number} \quad\quad (3 \text{ - } 43)$$

and the coverage is related to the amount sorbed, q, by

$$q = c\theta \quad\quad (3 \text{ - } 44)$$

The rate of sorption in terms of q is

$$dq/dt = \{mspN/(2\pi mkT)^{0.5} \exp[-E_o/kT]\} \exp[-aq/ckT] \quad\quad (3 \text{ - } 45)$$

which may be simplified to the Elovich equation:

$$dq/dt = a(T,p) \exp[-b(T)q] \quad\quad (3 \text{ - } 46)$$

For a nonuniform surface which may be subdivided into a finite number of domains, each a uniform surface, then the Elovich equation may be used to describe the sorption to these uniform domains. Here we may express the activation energy within a domain as

$$E_i = E_{o,i} + a_i\theta_i \quad\quad (3 \text{ - } 47)$$

where the subscript i denotes a particular domain. Summing the contributions over all the domains,

$$z = \Sigma z_i = p/(2\pi mkT)^{0.5} \Sigma s_i((1 - \theta_i) \exp[-(E_{o,i} + a_i\theta_i)/kt] \quad\quad (3 \text{ - } 48)$$

Now suppose $s_i = s$ for all i, and the summation is carried over many different domains such that n is large, then

$$z = sp/(2\pi mkT)^{0.5} \int_0^S ((1 - \theta)) \exp[-(E_o + aS)/kT]dS \quad\quad (3 \text{ - } 49)$$

Sorption Isotherms Derived from Statistical Mechanics

Thermodynamic properties can be developed, in principle, from a consideration of the statistical description of systems.[11] The power of the statistical mechanics approach is the direct connection between thermodynamics and statistical variables as

$$A = -kT \ln Q \quad\quad (3 \text{ - } 50)$$

where A is the Helmholtz free energy and Q is the *appropriate* partition function for the system of interest. For the present case we shall employ statistical methods to describe sorption of N_j identical molecules each with an average energy of $-E$ onto the N sorption sites with a stoichiometry of one molecule per site. The first step in this development is to construct the partition function from a description of the system. The grand partition function for such open systems (i.e., the interface is considered open) has three parts: (a) a degeneracy of the energy states, w; (b) a micro partition function to describe the fraction of the ensembles showing the required energy, E, at reduced temperatures, b; and (c) a function to describe the chemical potential, μ, of the adsorbed layer. Thus, the grand partition function is a summation over all the microstates:

$$q^* = \sum w \, exp[bN_iE] \, exp[bN_i\mu] \qquad (3 - 51)$$

For this system w is the number of ways of distributing N_j, identical molecules among N sites. This degeneracy is given by

$$w = N!/[(N_j)!(N - N_j)!] \qquad (3 - 52)$$

This expression may be explained as follows. The number of sites is N; $N!$ describes the number of different combinations for which N molecules may be distributed among N sites. However, the number of sorbing molecules, N_j, is assumed to be less than the number of sites, N; thus we need to consider only those combinations between N and $(N - N_j)$ occupied sites. So

$$w = (N)(N - 1)(N - 2) \, ... \, (N - N_j) \, ... \, (2)(1)/[(N - N_j) \, ... \, (2)(1)]$$
$$= N!/(N - N_j)! \qquad (3 - 53)$$

If N_j molecules are indistinguishable, the preceeding expression must be divided by $(N_j!)$. The expression for the grand canonical partition function for the present case becomes

$$q^* = \sum N!/[(N_j)!(N - N_j)!] \, \{exp[b(E + \mu)N_i]\} \qquad (3 - 54)$$

Added insight to this equation may be gained by expanding the summation over the N microstates as

$$q^* = N! \, exp[b(E + \mu)0]/0!N_i! + N! \, exp[b(E + \mu)1]/1!(N-1)! + ... \qquad (3 - 55)$$

$$q^* = 1 + N \, exp[b(E + \mu)] + N(N-1)/2 \, exp[b(E + \mu)2] + ... \qquad (3 - 56)$$

$$q^* = [1 + exp[b(E + \mu)]]^N \qquad (3 - 57)$$

Equation (3 - 57) is the *closed form* of the grand canonical partition function expressed as the summation in the prior equation. This partition function is particularly useful in determining the most probable number of molecules (N_p) sorbed to the surface. Recall the definition of the most probable number as

$$N_p = \Sigma N_i Pr(N_i) \qquad (3 - 58)$$

where $Pr(N_i)$ is the probability of N_i molecules being sorbed to the surface. This finite summation equation is related to the well-known integral expression for the first moment

$$n' = \int nP(n)dn \qquad (3 - 59)$$

where $P(n)$ is the probability of n having a value between n and $n + dn$. For the present case

$$Pr(N_i) = w \, exp[b(E + \mu)N_i]/q^* \qquad (3 - 60)$$

Thus, the mean value is just

$$N_p = \Sigma[N_i w \, exp[b(E + \mu)N_i]]/q^* \qquad (3 - 61)$$

Now examine the partial derivative of the grand canonical partition function as

$$\partial q^*/\partial \mu = \partial/\partial \mu \{\Sigma w \, exp[bN_i E] \, exp[bN_i \mu]\} \qquad (3 - 62)$$

$$\partial q^*/\partial \mu = b\Sigma w N_i \, exp[b(E + \mu)N_i] \qquad (3 - 63)$$

If we divide Eq. (3 - 63) by (bq^*) then

$$\{\partial q^*/\partial \mu\}/bq^* = (1/q^*) \Sigma w N_i \, exp[b(E + \mu)N_i] \qquad (3 - 64)$$

Upon elimination of the r.h.s. of Eq. (3 - 64) with its equal in Eq. (3 - 60) gives

$$(1/b)\partial ln[q^*]/\partial \mu = N_p \qquad (3 - 65)$$

The mean value, N_p, is

$$N_p = (1/b)\partial/\partial \mu \{ln[1 + exp[b(E + \mu)]]^N\} \qquad (3 - 66)$$

Let us evaluate the derivative as

$$ln q^* = N \, ln[1 + exp[b(E + \mu)]] \qquad (3 - 67)$$

$$\partial lnq'/\partial \mu = Nb \exp[b(E + \mu)]/\{1 + \exp[b(E + \mu)]\} \qquad (3 - 68)$$

$$(1/Nb)\partial lnq'/\partial \mu = \exp[b(E + \mu)]/\{1 + \exp[b(E + \mu)]\} \qquad (3 - 69)$$

$$N_i/N = \exp(bE) \exp(b\mu)/\{1 + \exp(bE) \exp(b\mu)\} \qquad (3 - 70)$$

where N_p/N is the most probable fractional coverage, θ. The chemical potential of the adsorbed layer being equal to the chemical potential of the gas phase can be calculated as follows

$$\exp(\mu/kT) = [h^2/(2\pi mkT)]^{3/2} p/kT \qquad (3 - 71)$$

assuming ideal gas behavior.[11] Thus the most probable fractional coverage is

$$\theta = \exp(E/kT)[h^2/2\pi mkT]^{3/2}(p/kT)/\{1 + \exp(E/kT) [h^2/2\pi kT]^{3/2}(p/kT)\} \qquad (3 - 72)$$

which is the Langmuir isotherm if $b = 1/kT$. Let us review the assumptions inherent to this development.

1. Gas phase is ideal, thus allowing μ to be calculated from Eq. (3 - 71).
2. Adsorption stoichiometry is one site/molecule; other stoichiometries will alter the expression for w.
3. Each site shows the same energy, E, where E is independent of the surface population (i.e., coverage, N_i, and so on).

The results from this statistical mechanical treatment are essentially the same as those obtained from the collision theory treatment.

Conclusions

These two derivations of adsorption isotherms from the kinetic and statistical treatments give valuable insight on how the surface structure influences the potential energy of the adsorbed species. Also, these models may be used in some cases to correlate sorption rate data. The models suggest a preliminary framework for proposing sorption and reaction mechanisms.

REFERENCES

1. Benson, S. W., The Foundations of Chemical Kinetics (New York: McGraw-Hill Book Co., 1960).
2. Present, R. D., Kinetic Theory of Gases (New York: McGraw-Hill Book Co., 1958).

3. Moore, W. J., *Physical Chemistry*, 2nd Edition (Englewood Cliffs, NJ: Prentice-Hall, Inc., 1955).
4. Brunauer, S., *The Adsorption of Gases and Vapours* (Princeton, NJ: University Press, 1943).
5. Murphy, W. R., A. K. Schwartz, T. F. Veerkamp, and T. W. Leland., *A Quantitative Characterization of the Electronic Effects of Chemisorption on Powdered Solids* (Houston, TX: Rice University, 1978).
6. Eischens, R. P., S. A. Francis, and W.A. Plisken, *J. Phys. Chem.*, **60**, (1956), p. 194.
7. Bell, A. T. and M. L. Hair, "Vibrational Spectroscopies for Adsorbed Species," *ACS Symposium Series*, **137**, (Washington, DC: American Chemical Society , 1980).
8. Amdurr, I. and G. Hammes, *Chemical Kinetics: Principles and Selected Topics* (New York: McGraw-Hill Book Co., 1966).
9. Blyholder, G. and M. C. Allen, *J. Am. Chem. Soc.*, **91**:12, (1969), p. 3158.
10. Miller, A. R., *Proc. Camb. Phil. Soc.*, **43**, (1947), p. 232.
11. Kubo, R., *Statistical Mechanics* (Amsterdam: North Holland Publishing Co. and New York: Interscience, 1965), p. 93.

PROBLEMS

1. The time for which an oxygen atom remains adsorbed to a tungsten surface is 0.36 s at 2548°K and 3.49 s at 2362°K. Find the activation energy and the pre-exponential factor for the desorption of oxygen from this surface (From *Physical Chemistry*, by P. W. Atkins. Copyright 1978. Reprinted with the permission of W. H. Freeman & Company.)

2. Nitrogen gas adsorbed on charcoal to the extent of 0.92 cc/g at 4.8 atm and 190°K, but at 250°K the same amount of sorption was achieved only when the pressure was increased to 32 atm. What is the molar enthalpy of adsorption for nitrogen on charcoal? (From *Physical Chemistry*, by P. W. Atkins. Copyright 1978. Reprinted with the permission of W. H. Freeman & Company.)

3. Calculate the time required for 10% of the sites on a tungsten {100} surface to be covered with nitrogen at 298°K when the pressure is 2 nano-Torr and the orientation factor is 0.55.

4. The chemisorption of hydrogen on manganese is activated, but only weakly so. Measurements have shown that it proceeds 25% faster at 1000°K than at 600°K. What is the activation energy for this chemisorption? (From *Physical Chemistry*, by P. W. Atkins. Copyright 1978. Reprinted with the permission of W. H. Freeman & Company.)

5. A mixture of gases A and B is in equilibrium with a solid surface. Show that if the Langmuir theory is used, then (assuming associative chemisorption)

$$\theta_a = K_a P_a / (1 + K_a P_a + K_b P_b) \qquad (3 \text{-} 73)$$

What is the corresponding expression for the fractional coverage of B?

7. The Langmuir isotherm is given by

$$\theta = Kp / (1 + Kp) \qquad (3 \text{-} 74)$$

for the single site, single adsorbate case; whereas the Temkin isotherm is given by

$$\theta = C_1 \ln(C_2 kp) \qquad (3 \text{-} 75)$$

and the Freundlich isotherm is given by

$$\theta = Cp^{1/n} \qquad (3 \text{-} 76)$$

Examine the following adsorption data of CO onto charcoal at $273°K$ to determine which of the three isotherms best describes these data.

P (Torr):	100	200	300	400	500	600	700
V (cc/g):	10.2	18.6	25.5	31.4	36.9	41.6	46.1

where V is the STP volume adsorbed per g of sorbent. (From *Physical Chemistry*, by P. W. Atkins. Copyright 1978. Reprinted with the permission of W. H. Freeman & Company.)

8. Given the hydrogen chemisorption data on an oxide at the two temperatures ($305°$ and $444°C$), calculate the Langmuir equilibrium adsorption constants and the monolayer coverage volumes at each temperature. What comments can you make regarding the isosteric heat of adsorption? Comment on the response of the monolayer volume with temperature. A model used often to correlate the adsorption equilibrium constants assumes

$$\ln K = -(\Delta G_{ads}) / RT \qquad (3 \text{-} 77)$$

Develop a relationship that allows a determination of the enthalpy and entropy of adsorption from a knowledge of the equilibrium adsorption constants at two different temperatures. Calculate the enthalpy and entropy of adsorption from these data.

$T = 305°C$		$T = 444°C$	
P (Torr)	V_{ads} (scc/g)	P (Torr)	V_{ads} (scc/g)
44	156.9	3	57.1
51	160.8	22	83.3
63	163.8	48	95.0
121	167.0	77	98.1
151	169.6	165	100.9
230	171.1		
269	171.6		

(Reprinted with permission from H. S. Taylor and A. T. Williamson, *J. Am. Chem. Soc.*, **53**, (1931), p. 2168. Copyright 1931 American Chemical Society.)

9. Derive the Elovich Equation and evaluate the constants for the chemisorption of hydrogen on Ni/alumina.

| Table 3 - 3 Hydrogen Adsorption on Ni/alumina | |
Time (min)	V_{ads} (scc/g)
0	0
1.0	8.89
1.5	8.93
3.0	9.01
5.9	9.08
10.0	9.13
14.5	9.18
20.4	9.20
28.8	9.24
59.0	9.32
120.0	9.39
148.0	9.41

(Reprinted with permission from S. Narayanan and L. M. Yeddanapalli, *J. Catal.*, **21**, (1971), p. 356.)

10. The adsorption of Ar on graphitized carbon films at the liquid nitrogen point (77.8°K) and the liquid oxygen point (90.1°K) has been reported as in Table 3 - 4. Fit the Langmuir Equation to these data and determine adsorption equilibrium coefficient, K, in the units of reciprocal Torr; the monolayer volume in STP cc/g; and the heat of sorption in kcal/mole.

11. Consider the Langmuir and Temkin isotherms for sorption as given here.

$$\theta/(1 - \theta) = [K_o \, exp(Q/RT)]P \qquad (3 - 78)$$

where

θ = fractional coverage
K_o = fitting constant, Torr^{-1}
P = adsorbate pressure, Torr
R = gas constant, cal/mole-°K
T = temperature, °K
Q = heat of adsorption
  = Q_o for Langmuir treatment
  = Q_o - $a\theta$ for Temkin treatment
a = fitting constant for the Temkin isotherm

Plot the fractional coverages versus adsorbate pressure for these two isotherms using the following constants:

K_o(Temkin) = 1.234 x 10^{-3} Torr^{-1}; a = 5000 cal/mole; Q_o = 5000 cal/mol

K_o(Langmuir) = 2.468 x 10^{-4} Torr^{-1}; Q_o = 5000 cal/mole; T = 503.2°K

Plot the Temkin isotherm once again, but using the temperature equal to 1006.4°K. Calculate the isosteric heats of sorption using these Temkin data *but* assuming the data following the Langmuir isotherms. Use the value of K_o(Temkin) that you used for the 503.2°K isotherm but corrected as follows:

$$K_o(T_1)/K_o(T_2) = (T_2/T_1)^{0.5} \qquad (3 - 79)$$

Table 3 - 4 Argon Adsorption on Graphitized Carbon

| Pressure (Torr) | V_{ads}/g catalyst (STP cc/g) | |
	77.8°K	90.1°K
0.05	0.74	
0.10	1.20	0.175
0.20	1.75	
0.30	2.07	
0.40	2.27	
0.50	2.42	0.714
0.70	2.60	
0.80	2.67	
0.90	2.72	
1.00	2.76	1.16
2.00		1.69
3.00		1.99
5.00		2.33
10.00		2.66

PART II

EXPERIMENTAL METHODS

The catalyst must be characterized such that "structure" may be related to performance. Here, we use the term "structure" to mean the physical properties of the catalyst and may include bulk as well as surface properties. The discussion of surface characterization is limited to a very few of the many tools available; however, the choice is made to include some of the more common techniques often encountered in industrial and academic laboratories. The discussion is divided between "classical" tools such as volumetric and gravimetric adsorption apparatus and ultrahigh vacuum equipment such as AES/XPS.

Characterization of the catalyst for performance is limited to reactivity and selectivity in chemical reactors. The types of reactors are discussed to relate the relative merits of reactor type. With such information, the student will be able to make a rational choice of reactor type in future studies.

CHAPTER 4

CLASSICAL METHODS

FOR SURFACE CHARACTERIZATION

In the preceding chapters, we have described the theories of surface/sorbate interactions without giving much attention to the actual mechanics for measurement of the same. In this chapter we present a brief overview of the classical, experimental methods for characterization of the surface; whereas the ultrahigh vacuum (UHV) surface science tools are discussed in Chapter 5. We begin with a discussion of a volumetric device to measure the total surface area of an adsorbent by physisorption of the appropriate adsorbate. This same device may be used to describe "active" surface area if a selective adsorbate is employed. The energy of the sorption process may be characterized by the vibrational spectroscopy of selected chemisorbed species combined with the temperature programmed desorption spectra of selectively chemisorbed species. Of course, calorimetry is the most direct method to determine the energy of sorption. The site stoichiometry and morphology of the sorbed layer are important to understanding the sorption process. Vibrational spectroscopy of sorbed molecules is a useful technique to describe the site stoichiometry/morphology.

Methods of Studying the Adsorption Phenomenon

Before we describe the procedures of adsorption, consider the basic approach. The adsorption process may be followed by
1. changes in the gas phase properties (pressure, and/or volume)
2. changes in the solid phase properties (electrical conductivity, magnetic properties, and sample weight
3. changes in the adsorbed phase properties (interaction of photons, electrons, ions, or atoms with the adsorbed phase)

If the changes in the gas phase properties are chosen then there is a choice of examining such changes in a static or continuous system. Consider now a mass ("standard" volume) balance on the adsorbing gas.

$$V_p = V_t + V_d - V_a \quad (static)$$

$$(4 - 1)$$

$$V_p = V_1 + V_d - V_a - V_2 \quad \text{(dynamic)} \qquad \text{(4 - 2)}$$

where

V_p = accumulation of gas above the adsorbent
V_1 = amount of gas injected into the apparatus
V_2 = amount of gas removed from the apparatus
V_a = amount of gas adsorbed
V_d = amount of gas desorbed

For the static experiment at steady state, the molecules are in dynamic equilibrium between surface and gas phase. The net adsorption is V_n:

$$V_n = V_a - V_d \qquad \text{(4 - 3)}$$

Volumetric Method

The basic idea is to estimate the moles of gas adsorbed from a knowledge of the system volume and the conditions within the system. The errors associated with determining the moles of gas sorbed are those in reading the temperature and pressure of the system and in measuring the system volumes. In the classical, static procedure, a sorbing gas is expanded from a burette of known volume at a given temperature and pressure (such that the initial number of moles may be calculated) into the adsorption cell (of known volume) containing the adsorbent. A molar balance on the gas phase is

$$N_a = N_i - N_f \qquad \text{(4 - 4)}$$

where

N_a = net number of moles adsorbed
N_i = number of moles of gas charged to apparatus
N_f = number of moles remaining in gas phase

The moles in the gas phase may be calculated from consideration of the system temperature, pressure, and volume, using the appropriate gas phase equation of state. A simple volumetric device (Fig. 4 - 1) is used here to describe the technique. The total volume of the burette system is adjusted by raising/lowering a mercury meniscus within the burette. Typically, the volumes (V_a, V_b, and so on) have been measured prior to the experiment and a suitable nonsorbing gas, such as helium, is used to determine the dead volume of the lines including the cell volume not occupied by the adsorbent. This volume is calculated as follows:

$$N_i = N_f$$

Assuming the gases behave according to the modified Ideal Gas Law, then

$$p_i(V_1 + V_2 + V_3 + V_4 + V_5)/RT =$$

$$p_f(V_1 + V_2 + V_3 + V_4 + V_5)/RT + p_f(V_{cell})/(RT_c z)$$ (4 - 5)

where

p_i = initial pressure
p_f = final pressure
T = room temperature
V_{cell} = cell volume-adsorbent volume
z = compressibility factor of gas at cell conditions
T_c = cell temperature

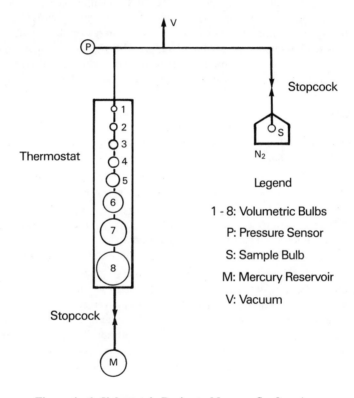

Figure 4 - 1 Volumetric Device to Measure Surface Area

A similar equation can be written for the experiment where the valve to the cell is closed so as to elucidate the dead volume, V_d. It is now possible to describe the moles of gas sorbed, N_{ads}, as a function of pressure:

$$N_{ads}(p) = N_i - N(p)$$ (4 - 6)

where

N_i = initial number of moles in gas phase

$N(p)$ = moles of gas present at pressure, p

N_{ads} = number of moles of gas adsorbed at pressure, p

Again, N_i and $N(p)$ are calculated from the data of T, p, and V.

Types of Isotherms

Brunauer[1] describes five distinct physisorption isotherms (see Fig. 4 - 2). The shapes of these isotherms suggest the physics of the adsorption process. For example the type I adsorption isotherm defines monolayer adsorption; whereas the type II isotherm indicates multilayer sorptions. The BET equation correctly models the behavior of type II isotherms. The type III isotherm describes very weak sorbent/sorbate interactions for which the cohesive energy of the liquid is stronger than the gas/solid interactions. Poehlein[2] reports the sorption of noble gases onto alkali metal superoxides to follow type III isotherms. Isotherms of type IV and V model the physisorption to catalysts showing the effects of capillary condensation.

BET Method

Brunauer, Emmett, and Teller[3] describe a technique to measure total surface area based on the multiple layer physisorption. The assumptions of the theory are

1. Fixed number of adsorption sites.
2. Second and subsequent layers sorb directly onto the first layer.
3. No lateral interactions between molecules absorbed within a layer.
4. Constant heat of adsorption within each layer.
5. Heat of adsorption in first layer is H; for all other layers heat of adsorption is equal to heat of liquefaction.
6. Infinite adsorption at the saturation conditions of the adsorbate.
7. Dynamic equilibrium between sorbed and gas phase.

The model of this physisorption is given in Figure 4 - 3. The amount of surface left uncovered is designated S_o, that covered by one layer of molecules is S_1, and so on. Postulate 7 expresses the concept of microreversibility to be applied to *each* layer. Thus, for a particular layer i, the rate of adsorption is just equal to the rate of desorption from the ith layer. Borrowing from the concepts of collision theory, we write these "equilibrium" adsorption/desorption relationships as

$$rate\ of\ adsorption\ =\ rate\ of\ desorption \qquad (4 - 7)$$

for the *ith* layer

$$(k_i)p(S_{i-1}) = (k_{-i})S_i \qquad (4 - 8)$$

Here, we express

p = pressure of the adsorbing gas

k_i, k_{-i} = forward and reverse rate constants

Equation (4 - 8) is rearranged as

$$p[S_{i-1}] = S_i(k_{-i}/k_i) \qquad (4 - 9)$$

$$p[S_{i-1}] = S_i(b_i/a_i) \, exp(E'/RT) \qquad (4 - 10)$$

where

b_i, a_i =   preexponential factors for reverse, forward rate constants;
$b_1/a_1 = g'$; $b_2/a_2 = b_3/a_3 = b_i/a_i = g$

E' = $(E_f - E_r)$ = difference in activation energies of the forward and reverse

= $-H'$ for the first layer ($i = 1$)

= $-H_v$, heat of liquefaction for $i \geq 2$

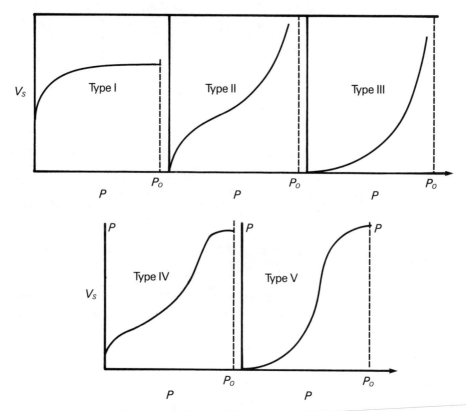

Figure 4 - 2   Types of Physisorption Isotherms

Reprinted with permission from S. Brunauer, et al., *J. Am. Chem. Soc.*, **52**, (1940), p. 1723. Copyright 1940, American Chemical Society.

The total surface area A is the sum of S_i over all n layers

$$A = \sum_{j=0}^{n} S_i \qquad (4 - 11)$$

The total volume of molecules adsorbed is the sum over n layers as

$$V_{ads} = \sum_{j=1}^{n} V_i \qquad (4 - 12)$$

$$V_{ads} = (V_m/A) \sum_{j=1}^{n} (jS_j) \qquad (4 - 13)$$

Now, ratio the volume adsorbed to the surface area of the solid to yield

$$V_{ads}/A = (V_m/A) \sum_{j=1}^{n} jS_j / \sum_{j=0}^{n} S_j \qquad (4 - 14)$$

or, upon rearrangement,

$$V_{ads}/V_m = \sum_{j=1}^{n} (jS_j) / \sum_{j=0}^{n} S_j \qquad (4 - 15)$$

Equation (4 - 10) may be rearranged to yield

$$S_1 = S_o p(a_1/b_1) \exp[(E_1 - E_o)/RT] \qquad (4 - 16)$$

$$S_1 = S_o(p/g^*) \exp(H^*/RT) \qquad (4 - 17)$$

$$S_2 = S_1(p/g) \exp[(E_2 - E_1)/RT] = S_1(p/g)\exp[H_v/RT] \qquad (4 - 18)$$

$$S_2 = S_o(p/g^*)(p/g) \exp[(H^* + H_v)/RT] \qquad (4 - 19)$$

$$S_3 = S_o(p/g)^3(g/g^*) \exp[(H^* + 2H_v)/RT] \qquad (4 - 20)$$

or

$$S_3 = S_o(p/g)^3 C \exp(3H_v/RT) = S_o[(p/g) \exp(H_v/RT)]^3 C \qquad (4 - 21)$$

$$S_3 = Cx^3 S_o \qquad (4 - 22)$$

where
$$C = (g/g^*) \exp[(H^* - H_v)/RT]$$
$$x = (p/g) \exp(H_v/RT)$$
Thus, a recursion formula may be established for S_j in terms of S_o

$$S_j = C(x^j)S_o \qquad (4 - 23)$$

to relate the volume adsorbed to the monolayer volume is as follows:

$$V_{ads}/V_m = C\sum_{j=1}^{n}(jx^j)/\{1 + C\sum_{j=1}^{n}x^j\} \qquad (4 - 24)$$

Gaseous Molecules

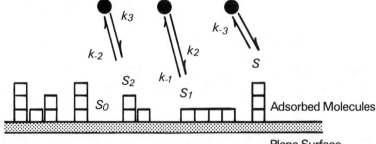

So Fraction of Surface Not Covered S_3 Fraction of Surface Covered With Three Molecules
S_1 Fraction of Surface Covered With One Molecule
S_2 Fraction of Surface Covered With Two Molecules

Figure 4 - 3 BET Physisorption Model

These finite series may be expressed in closed form as the following:

$$\sum_{j=1}^{n}x^j = (x - x^{(n+1)})/(1 - x) \qquad (4 - 25)$$

and

$$\sum_{j=1}^{n}(jx^j) = x\, d/dx \sum_{j=1}^{n}\{x^j\} \qquad (4 - 26)$$

so that

$$\sum_{j=1}^{n}(jx^j) = x\, d/dx\{(x - x^{(n+1)})/(1 - x)\} \qquad (4 - 27)$$

or

$$\sum_{j=1}^{n}(jx^{j}) = x/(1 - x)^2 \{[1 - (n + 1)x^n + nx^{(n+1)}]\} \qquad (4 - 28)$$

$$\sum_{j=1}^{n}(jx^{j}) = 0 x_0 + \sum_{j=1}^{n}(jx^{j}) \qquad (4 - 29)$$

Thus, the final expression is

$$V_{ads}/V_m = Cx/(1 - x)\{1 - (n + 1)x^n + nx^{(n+1)}\}/\{1 + (C - 1)x - Cx^{(n+1)}\} \qquad (4 - 30)$$

Examination of Eq. (4 - 30) reveals two cases for the value of V_{ads}/V_m as n approaches infinity.

Case I: $x = 1$

$$\lim_{n \to \infty}(V_{ads}/V_m) = C/0 \{1 - (n + 1) + n\}/\{1 + C - 1 - C\} \qquad (4 - 31)$$

This condition requires a repeated appeal to L'Hopital's rule to remove the singularity. Taking the derivative of the numerator and the denominator yields upon several applications

$$\lim_{n \to \infty}(V_{ads}/V_m) = C(n^2 + n)/[2(1 + Cn)] \qquad (4 - 32)$$

Equation (4 - 32) shows V_{ads}/V_m is unbounded at $x = 1$ when n increases to large values. This mathematical condition agrees with the physical condition of multilayer sorption when p/p_o is unity. Thus, we conclude

$$x = 1 \text{ when } p/p_o = 1$$

and recalling from the definition of x

$$x = (p/g) exp(H_v/RT)$$

then the ratio of x/x_o must be

$$(x/x_o) = p/p_o$$

The compelling conclusion from this development is that x equals p/p_o since x_o equals unity. For values of x greater than unity the sorption process degenerates into a *condensation* process for which the present model will not address. Thus, we restrict our attention to those values of x less than or equal to unity.

Equation (4 - 30) predicts type I isotherms for the condition where n equals unity. That is,

$$V_{ads}/V_m = Cx/(1 - x)\{1 - (1 + 1)x + x^{(1+1)}\}/\{1 + (C - 1)x - Cx^{(1+1)}\}$$

$$= Cx/(1 + Cx) \tag{4 - 33}$$

represents the Langmuir equation when x is p/p_o.

Case II: x less than 1

For this case, the limit as n increases without bound is given by

$$lim(V_{ads}/V_m) = \lim_{n \to \infty}\{Cx/(1 - x)[1 - (n + 1)x^n + nx^{(n+1)}]/[1 +(C - 1)x - Cx^{(n+1)}]\} \tag{4 - 34}$$

$$\lim_{n \to \infty}(V_{ads}/V_m) = Cx/[(1 - x)(1 + (C - 1)x)] \tag{4 - 35}$$

If we let $x = p/p_o$, then

$$V_{ads}/V_m = C(p/p_o)/\{(1 - p/p_o)(1 + (C - 1)p/p_o)\} \tag{4 - 36}$$

Equation (4 - 36) is the isotherm for fitting physisorption data when (p/p_o) is less than unity. Experience dictates that only those sorption data within the range of relative pressures (p/p_o) between 0.05 and 0.35 be used in extracting the model constants. For the application of determining sorbent surface area, one typically chooses the appropriate adsorbate for which an isotherm may be developed and is amenable to modeling. Type II isotherms are generally modeled by the BET equation. We present here the procedure for using the BET equation for some typical surface area determinations.

Linearizing the BET Equation

Equation (4 - 36) may be linearized as follows.

$$p/[V_{ads}(p_o - p)] = [(C - 1)/V_mC][p/p_o] + 1/(V_mC) \tag{4 - 37}$$

or

$$y = ax + b \tag{4 - 38}$$

Thus, the BET ordinate, y, is plotted versus the relative pressure, x, such that the parameters a and b are extracted. It is desired to determine the number of molecules, N_m, for which an ideal monolayer is formed. The total surface area is determined from the monolayer number density and the appropriate projected area per molecule. One may first extract the monolayer volume by adding a and b:

$$a + b = 1/V_mC + (C - 1)/V_mC = C/V_mC = 1/V_m \qquad (4 - 39)$$

and then taking the reciprocal,

$$V_m = 1/(a + b) \qquad (4 - 40)$$

This monolayer volume at standard conditions may be related to the surface area from a knowledge of the adsorbed phase properties. Realizing that V_m is an expression of the number of molecules required (N_m) to cover the surface by monolayer we may represent the surface area as

$$A = N_m a_p = (V_m/22400)N_o a_p \qquad (4 - 41)$$

where

A = specific surface area of the adsorbent; cm^2/g
V_m = monolayer volume in STP cc/g catalyst
a_p = projected area in cm^2/molecule

Brunauer suggests the projected area may be calculated from the following:[3]

$$a_p = 1.091(M/N_o\rho)^{(2/3)} \qquad (4 - 42)$$

where

M = formula weight
N_o = Avogadro's number
ρ = adsorbed phase density, gm/cc

In some applications the adsorbed phase density may be estimated using the liquid phase density at 1 atm. Some typical values of a_p are reported in Table 4 - 1.

Table 4 - 1
Projected Cross-Sectional Areas for Typical Adsorbates

Adsorbate	a_p; ang.^{2}/molecule
N_2	16.2
Kr	16.3
Ar	14.3

Example 1 Poehlein[2] reports the krypton physisorption on potassium superoxide for three samples (see Fig. 4 - 4) as volume adsorbed (STP) per gram of sorbent between

the relative pressures of 0.05 and 0.20. Samples 1 and 2 are sieved powders, 40-60 mesh; sample 3 is the unsieved powder. Comment on these data and report the most reliable estimate for the specific surface area.

Solution   Unlike some textbook problems this example is a slice of real life drama. The sorption isotherms showed anomalous behavior in the range of low relative pressures where the data extrapolated to zero pressure would indicate a *negative* volume adsorbed. This behavior is characteristic of type III isotherm behavior so we anticipated the linearized BET plot to be unusual.   Clearly, the low and high pressure data do not conform to the rest of the data. Meeting our expectation the BET plot (Fig. 4 - 5) yielded surface areas of 2.28 ± 1.64 m²/g (sample 1), 2.95 ± 2.65 m²/g (sample 2), and 1.26 ± 0.78 m²/g (sample 3) using all the data. If only the intermediate pressure data are used the reported area is 0.82 m²/g. These data all point to the conclusion that the BET equation cannot be used to correlate data of a Type III isotherm.   For type III isotherms Brunauer suggests the value of C in Eq. (4 - 36) is unity; thus

$$p/[V_{ads}(p_o - p)] = 1/V_m \qquad (4 - 43)$$

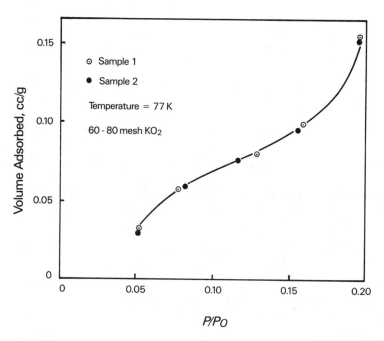

Figure 4 - 4 **Krypton Physisorption on Potassium Superoxide**

Reproduced with permission from J. O. Stull and M. G. White, *O. E. D., Vol. 11* (New York: American Society of Mechanical Engineers, 1985), p. 325.

Chapter 4 Classical Methods   73

The monolayer volume is just the volume adsorbed when $p = 0.5p_o$. For the present case, a linear extrapolation of the data from $p/p_o = 0.2$ to that at $p/p_o = 0.5$ results in the monolayer volume of 0.37 scc/g. Using a projected area of 16.3 angstrom^2/molecule for Kr gives a specific surface area of 1.62 m^2/g. We concluded that the surface area of these samples was between 1 - 2 m^2/g.

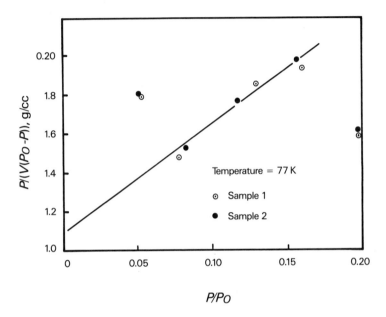

Figure 4 - 5 Linearized BET Plot of Kr Physisorption over Potassium Superoxide

Reproduced with permission from J. O. Stull and M. G. White, *O. E. D., Vol. 11* (New York: American Society of Mechanical Engineers, 1985), p. 325.

Capillary Restrictions -- Types IV and V Isotherms

For the very high surface area sorbents such as the molecular sieves and activated charcoals, a significant portion of the available surface area resides in small pores where the physical dimensions of these same may limit the maximum number of molecular layers which can be sorbed. For these cases, Brunauer[1] suggests the following equation:

$$V_{ads}/V_m = x/(1-x) + \{2(c-1)x + 2(c-1)^2x^2 + [nc^2+nh-2nc-(nc)^2]x^n +$$
$$[2c + (nc)^2 + 2nc-2c^2-nc^2-2h-2nh]x^{n+1} + [nh + 2h]x^{n+2}\}/$$
$$2\{1 + 2(c-1)x + (c-1)^2x^2 + [c^2+h-2c-nc^2]x^n + [nc^2 + 2c-2c^2-2h]x^{n+1} + hx^{n+2}\} \qquad (4 - 44)$$

where

$$h = [nc^2 - c^2 + 2c]g'$$
$$g' = exp[Q/RT]$$
$$Q = \text{heat of liquefaction}$$
$$c = exp[(E_1 - E_2)/RT]$$
$$x = p/p_o$$
$$n = \text{average number of layers to fill a capillary}$$

Brunauer[1] makes the following suggestions to fit Eq. (4 - 44) to Types IV and V isotherms:

1. For small values of x, p/p_o, the nth and higher powers of x may be discarded for which Eq. (4 - 44) reduces to the BET isotherm, Eq. (4 - 36). From these data of small p/p_o one may obtain V_m and C. The constants n and h may be determined from the following approximation to Eq. (4 - 44):

$$V_{ads}/V_m = 0.5n - [(n - 2)c]/[h^2(1 - x)^3 x^{(n-2)}]x^n - n(x)^{(n-1)} + nx^{(n-2)} \qquad (4 - 45)$$

2. Equation (4 - 45) is valid for p/p_o near unity for which $h \gg c^2$ and $c \gg 1$. A series expansion of Eq. (4 - 45) is useful in the evaluation of the constants.

Some final comments on this development are necessary:

1. It is assumed the capillaries are of uniform diameter; whereas most adsorbents will show a distribution of pore diameters. The characteristic capillary size, represented by n molecular adsorbent diameters, is an average size of the pore size distribution.

2. The capillary walls were assumed to be parallel when in fact there are data to show the capillary walls of actual adsorbents are not parallel.

3. Only those capillaries which are open at each end were modeled by this development; dead end pores were not considered.

Pore Volume Distribution from Adsorption-Desorption Hysteresis

The reversible physical adsorption of vapors onto high surface area adsorbents sometimes show a hysteresis near relative pressures of unity (see Fig. 4 - 6). One explanation of this hysteresis arises from the condensation of liquid in the capillaries.[4a, 4b, 4c, and 4d]

Consider the process of condensation in an array of cylindrical pores containing a distribution of pore radii and lengths with some open at both ends or closed at one end (Fig. 4 - 7). At the onset of adsorption, the walls of the pores will be covered by a layer of adsorbate of thickness, t. The relationship between the saturation pressure of the curved interface to that of the flat liquid interface is given by the Kelvin equation,

$$\ln(p_o/p) = 2\sigma V \cos\theta/[R_m RT] \qquad (4 - 46)$$

where

$2/R_m \quad = 1/R_1 + 1/R_2$

$R_m \quad$ = mean radius of curvature

R_1, R_2 = radii of curvature for the meniscus

$p_o, p \quad$ = saturation pressures for flat, curved interfaces

$\sigma \qquad$ = surface tension, 8.85 ergs/cm^2 for liquid N_2 at 77.4°K

$V \qquad$ = adsorbate liquid molar volume, 36.5 ml/mole for liquid N_2 at 77.4°K

$R \qquad$ = gas constant, 8.314 x 10^7 ergs/mole-$^\circ$K

$T \qquad$ = absolute temperature, $^\circ$K

$\theta \qquad$ = contact angle of meniscus with solid, assumed to be zero

In the present case, R_m is the mean radius of curvature for a closed-end pore showing two radii of curvatures. Condensation will occur to "bridge" the pore when $R_m = R_1$ = R_2 and the Kelvin equation relates the adsorbate pressure to the value of R_m at this condition. However, for an open-ended pore condensation will occur when $R_m = 2R_1$. The Kelvin equation may be rearranged to account for these cases as follows.

$$R_1 = 2\sigma V \cos\theta/[(1 + F)RT \ln(p_o/p)] = (r - t) \qquad (4 - 47)$$

In this equation, F is the fraction of pores open at both ends and is equal to zero for a sample having only closed-end pores. The radius R_1 may be related to the radius of the pore, r, and the thickness of the film, t, as shown in Eq. (4 - 47).

For the desorption branch, all the pores are filled with adsorbate; thus, F is identically equal to zero. The incremental desorption of adsorbate from the open- and closed-end pores is modeled in Figures 4 - 7c and 4 - 7d. Notice, the open-end pore loses adsorbate from both ends; whereas the closed-end pore loses adsorbate from only one end. Generally, the desorption branch of the isotherm is used to relate the incremental amount of adsorbate removed for an incremental lowering of the pressure above the sample. At each pressure, the Kelvin equation gives the average radius of the pores suffering the removal of adsorbate, Eq. (4 - 47).

The thickness of the adsorbed layer, t, also depends upon the reciprocal of the relative pressure, p_o/p as

$$t = A[\ln(p_o/p)]^{(-1/n)} \qquad (4 - 48)$$

such that

$$r = 2\sigma V \cos\theta/[(1 + F)RT \ln(p_o/p)] + A[\ln(p_o/p)]^{(-1/n)} \qquad (4 - 49)$$

Equation (4 - 49) relates the pore radius to the relative pressure, the properties of the adsorbate (V and σ), and the adsorbate/adsorbent contact angles. Thus knowing the relative pressure one knows the radius of the pores that *remain* filled with adsorbate on the *desorption* branch of the isotherm.

Consider now a typical experiment to determine the pore size distribution of a sorbent. The beginning of the experiment has the pores initially filled with $p/p_o = 1$. The pressure is lowered slightly to a new value of p/p_o and the volume desorbed, ΔV, is calculated from the following equation.

$$\Delta V = V_t - V \qquad\qquad (4 - 50)$$

where

$$V_t = \text{volume adsorbed at } p/p_o = 1$$
$$V = \text{volume adsorbed at new } p/p_o$$

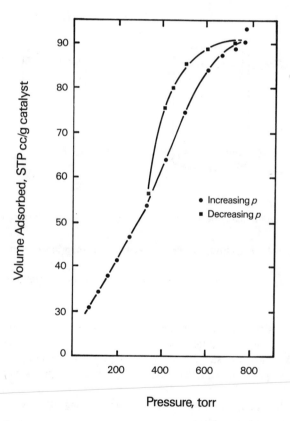

Pressure, torr

Figure 4 - 6   Adsorption-Desorption Hysteresis for Nitrogen on Silica-Alumina

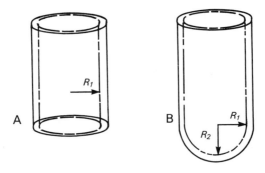

Adsorption in an ideal pore open at both ends and in an ideal pore closed at one end.

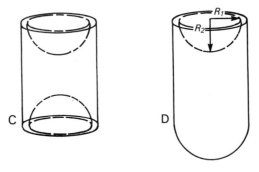

Desorption in an ideal pore open at both ends and in an ideal pore closed at one end.

Figure 4 - 7 Cylindrical Pore Models for Adsorption-Desorption

Used with permission from C. S. O'Neil, *American Laboratory*, June, (1985), p. 41.

For this new relative pressure, the radius of the pores remaining full is determined from Eq. (4 - 49) and the difference in radii for these two relative pressures is calculated as Δr. The pore size distribution is represented by the derivative, dV/dr, as a function of the pore radius, r. The derivative volume is approximated by the ratio $\Delta V/\Delta r$.

One of the common adsorbates for determining pore size distribution is nitrogen at $77°K$. For this case,

1. The contact angle is zero thus $\cos\theta = 1$

Chapter 4 Classical Methods

2. Wheeler[5] relates the sorbate phase thickness in angstroms as

$$t = 7.34(lnp_o/p)^{-0.333}$$ (4 - 51)

3. Hightower[6] reports the coefficient of $[ln(p_o/p)]$ as 8.86 using an exponent of -0.333
4. A modified Halsey-type equation gives t as [4c]

$$t = 3.54[-5/ln(p/p_o)]^{1/3}$$ (4 - 52)

5. A semiempirical relationship by Mingle and Smith[4d] gives t as

$$t = M[N/ln(p/p_o)]^{(1/L)}$$ (4 - 53)

where

$M = 4.132(A)$
$N = \ln[(1 - C^{0.5})/(1 - C)]$
$L = 1.42 [1 + 0.00212C]$
$A = 1.26(R)^{-0.085}$
$R = 2V_g/S$, average pore size
$V_g = $ liquid equivalent of maximum volume adsorbed
$S = $ BET surface area, m^2/g catalyst
$C = $ BET constant, C

Example 2   Consider the sorption data of N_2 at $77°K$ on a silica-alumina cracking catalyst (0.378 g). Develop the volume distribution for this catalyst.

Solution

The processed data of $(dV/dlnr)$ versus average pore radius are given (Fig. 4 - 8) for the adsorption and desorption isotherms. It is immediately apparent that the data do not describe a common pore size distribution (PSD) curve. Since both curves were obtained for nitrogen sorption to the same catalyst sample, one must conclude the differences in the curves are an artifact of the data reduction technique. O'Neil[4c] describes several cases of data reduction procedures used to characterize alumina (N10131). He showed the effects upon the PSD of different correlating equations used for thickness, t, of the sorbed layer. The PSD generated from the Halsey-type equations, Eq. (4 - 52), with $F = 0$ showed different distributions as in Figure 4 - 9. Better agreement of the PSD generated from the adsorption-desorption branches were possible if the Halsey-type equations were used having $F = 0.5$, but the BET surface area calculated from the adsorption branch exceeded that calculated from the desorption branch. The best results of PSD and BET surface area were obtained using the Mingle and Smith equation for the adsorbed layer thickness, t (Fig. 4 - 9).

Nitrogen Sorption Data for Example 2

Desorption Isotherm		Adsorption Isotherm	
V (scc/g)	P (torr)	V (scc/g)	P (torr)
93.45	758.9	30.80	64.8
90.13	703.9	38.09	145.3
88.25	580.0	41.69	186.5
85.54	490.6	46.89	246.8
79.88	427.9	53.89	318.3
75.10	397.5	63.62	405.2
56.29	332.9	74.56	493.9
		84.01	582.8
		87.56	647.8
		89.14	710.0
		90.45	744.8
		93.45	758.9

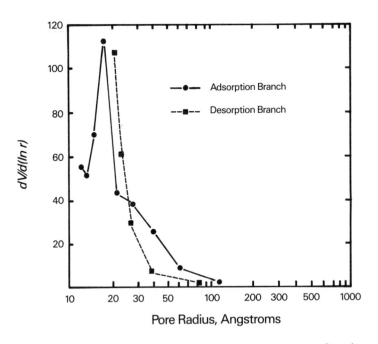

Figure 4 - 8 Pore Volume Distribution from Data of Nitrogen Sorption

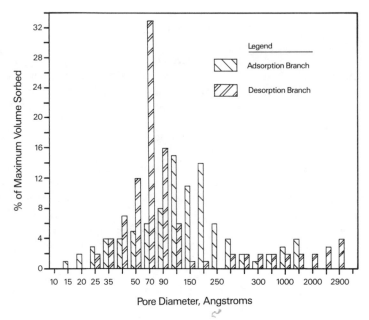

Figure 4 - 9 Pore Volume Distribution

Used with permission from C. S. O'Neil, *American Laboratory*, June, (1985), p. 41.

Pore Volume Distribution from Mercury Porosimetry

An alternate procedure for describing the pore volume distribution involves the mercury intrusion technique. The underlying principle of this technique is to relate the force necessary to "intrude" a nonwetting liquid, such as mercury, to the average radius of the pores which are filled. The number of these pores is related to the volume (called penetration volume) of mercury to fill pores of a certain average diameter.

The force required to fill the pores ($\pi r^2 \Delta p$) is set equal to the opposing surface tension force to impede such intrusion ($-2\pi r\sigma cos\theta$). This force balance on the intruding liquid gives an equation which relates the increase in pressure (Δp) to the radius of the smallest pores intruded.[7]

$$\pi r^2 \Delta p = -2\pi r\sigma \, cos\theta \qquad (4 - 54)$$

Solving for the pressure yields this familar relation for the surface tension, contact angle, and radius of curvature,

$$\Delta p = -2\sigma cos\theta / r \qquad (4 - 55)$$

where

Δp = pressure
σ = surface tension
r = radius of the pore
θ = contact angle

Usually the working fluid is mercury for which σ = 480 dynes/cm and θ = 140°. Equation (4 - 55) may be rearranged to give r (in angstroms) as p is given in atm.

$$r = 72,500/\Delta p \qquad (4 - 56)$$

This equation shows a radius of 21 angstroms for a pressure of 50,000 psia. This pressure represents a *maximum* useful limit as there is some question as to how the solid may respond (i.e., restructuring of the pores) to pressures in excess of 50,000 psia.

Typical data of pore volume versus pressure are shown in Figure 4 - 10 for three catalysts. Notice on this semilog plot the cumulative penetration volume is presented versus the intrusion pressure from which one may calculate the average pore radius.

The apparatus may consist of a calibrated piston pump, such as a Ruska Instruments pump, connected to a thermostatted high pressure cell and a pressure gauge. The zero penetration volume is noted when the pressure just begins to rise. Usually these instruments are automated such that the volume and pressure signals drive an X-Y recorder or the data are collected on small computer for post-run processing and manipulation.

The calculation of pore sizes from data of pore volume and specific surface area requires a model for the pore network. The level of model sophistication is attended by accuracy of the results and by effort required to extract the model parameters.

Pore Size Distributions

The most naive model assumes the pores to be uniform cylinders of radius, r, and length, l, which do *not* interconnect. From this simple model, a single parameter expresses a *gross* estimate of characteristic pore dimensions. Here, the pore volume/surface area measurements, V_p/S, are related to the assembly of N cylindrical pores having radius, r, and length, l, as

$$V_p/S = N(\pi r^2)l/N(2\pi rl) \qquad (4 - 57)$$

or

$$r = 2V_p/S \qquad (4 - 58)$$

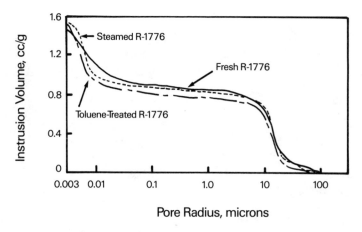

Pore Radius, microns

Figure 4 - 10 Cumulative Pore Volume versus Pore Radius for Silica-Alumina

Evidence from electron microscopy show the parallel pore, nonintersecting model is indeed the exception to the general rule. Studies[8] relate the measurables of pore volume and specific surface area to characteristic pore size, r, as

$$r = 2V_p/gS \qquad (4 - 59)$$

where g is given in Table 4 - 2. These models define a spectrum of characteristic pore dimensions which encompass the dimensions of the actual pores. However, it must be remembered *these* models all predict a uniform size and shape of the pores. Where the pore volume distribution data clearly show a bidisperse system of pore diameter (e.g., micropores and macropores), it may be possible to apply these simple models to each distribution such that characteristic dimensions may be derived for each distribution. Consider Figure 4 - 11 showing a bidisperse pore volume distribution. Clearly there exists local maxima in the microporous and macroporous regions for the sample. We may *define* two populations of pore sizes as follows: micropores, r less than 100 angstroms, and macropores, r greater than 100 angstroms. This decision is predicated on the shape of the pore volume distribution which shows two distinct peaks centered at 50 to 60 angstroms and 10^5 angstroms *and* a local minimum (nearly zero value of $dV/dlog\ r$) at $r = 10^2$ angstroms. We may apply the models to each distribution as

$$r_{micro} = (2/g)(V_p/S)_{micro} \qquad (4 - 60)$$

$$r_{macro} = (2/g)(V_p/S)_{macro} \qquad (4 - 61)$$

Table 4 - 2 Characteristic Shape Factors		
Structure	Length Parameter, r	Value of g
Nonintersecting Cylindrical Capillaries	Capillary Radius	1
Parallel-sided Fissures	Width of Gaps	1
Nonintersecting Close-packed Cylindrical Rods	Radius of Rods	0.104
Cubic Close-packing of Spheres	Radius of Sphere	0.613
Rhombohedral Packing of Spheres	Radius of Sphere	0.229
Orthorhombic Packing of Spheres	Radius of Sphere	0.433

Used with permission from J. M. Thomas and W. J. Thomas, *Introduction to the Principles of Heterogeneous Catalysis* (New York: Academic Press, Inc., 1967).

The pore volumes for the two distributions may be determined directly from the data of penetration volume (Fig. 4 - 10), or by integrating the data of Figure 4 - 11 between the appropriate limits. The measurement of the surface area for the two pore size groups represents a challenge. The usual BET method utilizes *adsorption* (increasing pressure) data to generate the volume of a monolayer. The Kelvin equation *cannot* be used to relate relative pressure (p/p_o) to the average pore size, as these pores are *filling* with adsorbate. Thus, the cumulative surface area as a function of pore radius cannot be determined. Consider now an application of the pore size distribution (PSD) method to characterize catalyst pretreatments.

The three curves of PSD in Figure 4 - 11 represent three silica-alumina catalysts: a fresh catalyst (R-1776), a fresh catalyst treated with toluene and calcined at $700°C$ with the hydrocarbon in the pores, initially (Toluene-Treated), and a fresh catalyst that was steamed-deactivated at $760°C$ for three hours (Steam-Deactivated). These catalysts showed BET surface areas of 342, 281, and 262 m^2/g catalyst as determined by N_2 physisorption at $77.4°K$. The mercury PSD results clearly show an effect of the physical treatments upon the microporous distribution of pore diameters while showing little or no effect upon the macroporous distribution of characteristic sizes. The "macropores" correspond to the size of the intersticial voids of a packed bed showing particle sizes of 60 to 80 microns. These PSD data together with the BET surface area data strongly suggest the toluene and steam treatments cause the micropores to collapse thus decreasing the total BET surface area and shifting the PSD to larger size pores.

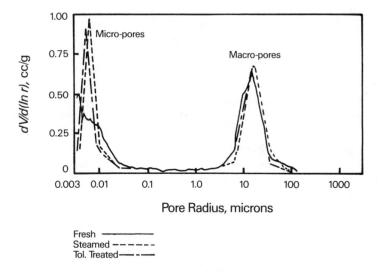

Figure 4 - 11   Pore Volume Distribution versus Pore Radius for Silica-Alumina

Selective Chemisorption

The earlier discussion on the. distinction between physisorption and chemisorption suggests a way to describe the surface using selective chemisorption agents. That is to say, the site density may be determined using a chemisorbing agent that is selective to only one type of site. The apparatus used for physisorption may be altered slightly for this purpose.

Table 4 - 3 gives a brief summary of the uses for some selectively chemisorbing gases. It must be remembered that the surface states of these solids may not necessarily correspond to the bulk phase and may be altered by the chemisorption process. For example, the chemistry of transition metals show the cupric ion will *not* form ligands with CO; however, some preparations of CuO will chemisorb CO from the gas phase. A closer examination shows these samples to contain cuprous ions in the surface region. Cuprous ions, shown by the chemistry of transition metals, will form ligands with CO.

Table 4 - 3
Selective Chemisorbing Gases

Metal/Oxide	N_2	H_2	O_2	CO	C_2H_4	C_2H_2
W,Mo,Zr,Fe	+	+	+	+	+	+
Ni,Pt,Rh,Pd	-	+	+	+	+	+
Cu,Al	-	-	+	+	+	+
Zn,Cd,Sn,Sb,Ag	-	-	+	-	-	-
Au	-	-	-	+	+	+
CuO		+	-		+	+
Cu_2O		+	+		+	+
SnO			-			

[+ = will chemisorb; - = will not chemisorb]

Selective Chemisorption to Metals

It has been shown that certain gases mentioned in Table 3 provide a sound basis for measuring the "active" site density for chemical reactions. For example, hydrogen has been used to describe the metal surface area of Pt supported on alumina catalyst. Two facts were established that allowed such a description:

1. H_2 sorbed *only* to the Pt and not onto the alumina surface.
2. The stoichiometry of H/Pt was well established as

$$1/2\ H_2(g)\ +\ Pt(surface) \rightarrow (Pt\text{-}H)surface \qquad (4 - 62)$$

Equation (4 - 62) implies that none of the H atoms penetrates into the bulk Pt metal. Thus, an experiment where H_2 gas dissociates on the surface to form up to a monolayer of Pt-H species allows the metal surface area to be calculated from the number of *exposed* Pt atoms.

The experimental conditions for this chemisorption experiment involve pressures to several hundred Torr and temperatures from room conditions to $200°C$. There remains the *very* remote possibility of weak chemisorption that is, reversible to evacuation which may cloud the results. This weak sorption problem can be solved by proper experimental technique. The experiment is to measure the volume adsorbed versus time for *two* successive exposures to H_2 separated by a mild evacuation treatment. The short evacuation serves to remove only the weakly sorbed H_2 which will be replaced upon the second H_2 exposure. The difference in the volumes adsorbed between the first and second exposures is defined by the strongly chemisorbed H_2 (see Fig. 4 - 12). Here, V_m is described by the value of the curve a-b at the limit of zero pressure. The stoichiometry of the chemisorption allows the measurement of V_m to infer the number of surface Pt atoms. Following Boudart,[9]

$$(V_m/22400)(2H\ atoms/molecule)(1Pt/H)N_o = No.\ of\ Pt\ atoms\ exposed/g \qquad (4 - 63)$$

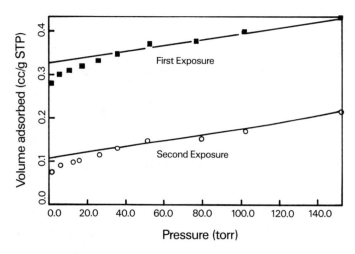

Figure 4 - 12   Selective Chemisorption of Hydrogen to Supported Pt Catalysts

Clearly the metal surface area in m^2/g may be calculated from Eq. (4 - 63) and a knowledge of the exposed metal crystal structure and lattice parameters.

Metal Dispersion

The number of exposed metal atoms divided by the number of total metal atoms of the same kind in the catalyst sample times 100 is called the percent dispersion. The dispersion is an indication of the efficiency of the impregnation procedure to "disperse every metal atom" upon the support and thus have *all* the metal atoms available for reaction. This goal of atomically dispersed metal may not be realized for all cases, and thus the metal may reside as small clusters for which some of the "interior" atoms are unavailable for reaction. Selective hydrogen chemisorption is often used to count the number of exposed metal atoms. However, there have been some questions raised on the validity of the Pt-H stoichiometry when the surface may be oxidized.   Mears and Hansford[10] suggest one possible alternative when the surface is PtO/support.

$$Pt\text{-}O \ + \ H\text{-}H \rightarrow H\text{-}Pt\text{-}H \ + \ H\text{-}O\text{-}H/support \qquad (4 \text{ - } 64)$$

Such questions of stoichiometry usually arise for samples which show *calculated* dispersions greater than 100%. Such is the case when CO is used to determine dispersion on some metals. It is known from IR data that the surface stoichiometry of CO to certain metals actually changes with coverage as shown in Figure 4 - 13. The problems of stoichiometry are obvious when the percent dispersions are greater than 100%, but one may ask if there is a stoichiometry problem for a sample that shows a percent dispersion of 90%. In this case one must calculate the average crystallite (or cluster) size and compare with an independent technique such transmission electron microscopy (TEM) or X-ray line broadening.

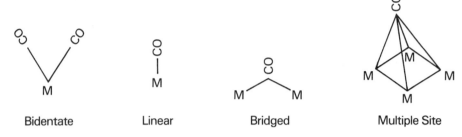

| Bidentate | Linear | Bridged | Multiple Site |

Figure 4 - 13   Surface Stoichiometries of CO to Metal

Crystallite Size

These same measurements of V_m may be used to calculate cluster size if one assumes a geometric model for the cluster and a model for the crystallographic planes presented to the surface. For Pt, the lattice parameters of two, low-index planes are given as follows: {110} plane shows 0.93×10^{15} atoms/cm^2; {100} plane shows 1.31×10^{15} atoms/cm^2. In general these clusters will present *many* different planes to the surface. By convention the combination of these two planes has been adopted as the "representative" crystallographic planes. This "average" plane shows an area of 9.2 angstrom^2/surface metal atom of Pt. Thus the dispersion data, together with the loading of metal on catalyst, may be used to calculate the exposed, metal surface area as follows:

$$Pt\ metal\ surface\ area\ =\ \{fractional\ dispersion\}\{N\}\{9.2\ angstrom^2/atom\}$$

where N is the number of Pt atoms in the sample. For other metals, the molecular weight, M, and the presented area will be different.

The cluster geometric model attempts to reproduce the actual cluster shapes. Electron microscope data show these clusters to be roughly spherical with some flat sides. Two models immediately come to mind: spheres and cubes. Let us consider the cube model. The *exposed* faces of a cube of size d affixed to a flat surface number five each of area d^2. Thus

$$exposed\ area\ =\ 5d^2 \qquad (4 - 65)$$

$$mass\ of\ cluster\ =\ \rho(d^3) \qquad (4 - 66)$$

for which the exposed area/g-cluster metal is given by

$$surface/g\ metal\ =\ 5/d\rho \qquad (4 - 67)$$

The surface area/g-catalyst is given by

$$surface/g\text{-}cat = \{surface/g\text{-}metal\}\{g\text{-}metal/g\text{-}catalyst\} \qquad (4 - 68)$$

The surface/g-metal becomes for platinum ($M = 195$ g/gmole)

$$S/d\rho = (\%D/100)(6.02 \times 10^{23}/195)(9.2 \times 10^{-16}cm^2/atom)$$

or

$$\rho d = 1.82 \times 10^{-4}/(\%D) \qquad (4 - 69)$$

The density used in Eq. (4 - 69) can be difficult to describe when the clusters are small. One solution is to assume the ρ is the same of that of bulk Pt (21.45 g/cc). Thus at 100% dispersion ($\%D = 100$) the characteristic dimension for the cube model is

$$d = 8.21 \times 10^{-8}\ cm \qquad (4 - 70)$$

Selective Chemisorption to Oxides

It appears reasonable to extend the techniques of selective chemisorption on metals to characterize metal oxides; however, the picture for the surface morphology of oxides is considerably more complex than that of metals. In general, the surface of oxides may contain atoms other than metal ion and oxidic oxygen, such as hydroxyl. The surface may suffer from defects such as coordinately unsaturated sites (CUS) in the metal ions, anion defects, altervalent impurity ion defects, and of course as with metals, lattice defects.

One family of oxides, the solid acids, has been studied in great detail in an attempt to relate the results of selective chemisorption experiments to the reactivity of the catalysts for model reactions. Benesi and Winquist[11] summarize the work of many researchers in this quest. Depending upon the system, the site density determined by the chemisorption and the model reaction agreed within one order of magnitude or, in the worst cases, the site densities determined by the base titrations were 1000-fold greater than that determined by the activity of the model reaction. Site energies have been determined by various indicator methods, with the nonaqueous methods showing the best agreement between the energy of sorption and reactivity of the sample. The aqueous solvents are known to react with the acid surface, thus tainting the results of the indicator. As with traditional indicator methods for determining acid strength of liquid acids, the solid acid is titrated with base to a certain endpoint as indicated by the color change in the indicator. The method has at least two deficiencies:

    1. The number of suitable indicators is limited.

2. Color changes of some indicators may be produced by processes other than simple proton addition.

The titration of surface acidity by gaseous bases has enjoyed limited success. Usually two approaches are employed to describe the acid site density and energetics of chemisorption. In the first case the acid is exposed to a gaseous base at a prescribed temperature, partial pressure of base, and duration. The solid is purged with inert gas for a duration usually at the same temperature, and the net pickup of base is recorded as the amount chemisorbed. This procedure is repeated for several different partial pressures of base at the same temperature to develop the experimental isotherm. Figure 4 - 14 shows an ammonia isotherm at 56°C for sorption to an amorphous silica-alumina cracking catalyst from Rosenthal.[12] The ultimate pickup of ammonia is determined by inspection (asymptotic pickup) or by extracting the model constant corresponding to monolayer coverage by a least-squares curve-fit of the data to a linearized sorption model (see Chapter 3).

Ammonia has been criticized as a poor chemisorbate in that the site density overestimates the cracking activity realized by model reaction studies.[13] These gaseous bases "count" sites which are sterically accessible and have acidities in the range titrated by the base. The small dimensions of ammonia allow it to access sites which are sterically inaccessible to larger bases such as pyridine. Moreover, the strong basicity of ammonia enables strong chemisorption to sites which are weak acids; whereas a weaker base such as pyridine will not be retained on such sites

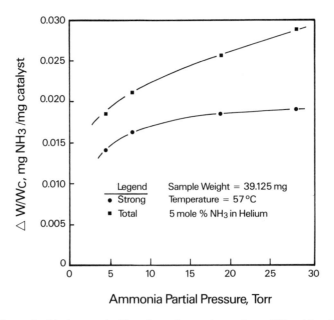

Figure 4 - 14 Ammonia Chemisorption to Amorphous Silica-Alumina

during the inert purge. The bases count the sites just as one would expect. It is left to the scientist to choose the proper base(s) for specifying acid site densities having the range of acidities deemed active for the model reaction of interest.

The gaseous bases mentioned, ammonia and pyridine, titrate Lewis and Bronsted acidity; thus, they are not selective as to type of acidity. The individual proportions of site density have been elucidated in some studies using a selecitve surface reactant such as hexamethyldisilazane (HMDS) or perylene. HMDS is a reputed reactant for silanols, hence Bronsted acids; whereas perylene reacts with the CUS of metal ions, hence Lewis sites.[12, 14] We[12] have used HMDS in conjunction with ammonia chemisorption to describe the fraction of acid sites on an amorphous silica-alumina (20 wt% Al_2O_3). We used IR spectroscopy of the chemisorbed pyridine to deduce the relative populations of sites. This sample showed 36% of the sites were Bronsted and the balance were Lewis acid sites.

Energy of Sorption

The earlier discussions of the potential energy associated with the sorption process now prompts a description of the experimental techniques to characterize the heat effects associated with sorptions. The most obvious way to describe the heat effects is through calorimetry; however, the method is not without its limitations. Heat fluxes are determined from a knowledge of temperature gradients through materials of known thermal properties. The reader is referred to reviews[15, 16, 17] describing these calorimetric procedures. Indirect methods for *estimating* the heats of sorptions range from *qualitative* determinations using vibrational spectroscopy to semiquantitative using temperature programmed desorption.

Isosteric Heat of Adsorption

The energy of sorption may be estimated from the equilibrium adsorption at several temperatures according to the theory given in Chapter 3, Eq. (3 - 18). Here the isosteric (constant fractional surface coverage) heat of adsorption is determined from the partial pressure of adsorbate required to attain a prescribed coverage at temperatures T_1 and T_2. Figure 4 - 15 is a plot of isosteric heat of adsorption data developed from ammonia chemisorption at $40°$, $80°$, and $125°C$ for three metal oxide complexes supported on Cab-O-Sil (28 wt% complex). These data show the surfaces of all three model catalysts are "ideal".[18] The results are not as typical as most commercial catalysts which show decreasing heats of adsorption with increasing fractional coverages.

Vibrational Spectroscopy

Consider a molecule such as CO sorbed onto the surface of a metal. We seek to describe the energy of the surface metal atom-carbon atom since it is known that CO bonds via the carbon atom.[19] In one sense, the sorption of CO onto some metals may be considered as the reaction to form a metal carbonyl. As such, the strength of the metal-carbon bond should be amenable to characterization by vibrational spectroscopy. The energy of these metal-carbon bonds sets the vibrational frequency in the near infrared range, where the relaxations tend to be broad and weak for solids. Subtle differences in metal-carbon bond strengths are not obvious using this part of the vibrational spectrum. All is not lost, however, for the energy of the metal-carbon may be inferred from the energy of the bond next removed from the surface: the C-O bond. This bond is IR active and produces "relatively" sharp relaxations. Additionally, the energy of the CO bond is very sensitive to the type of bonding that exists between the metal and the carbon (see Fig. 4 - 13). These drawings depict the essential features of the types of CO interactions with the surface metal atoms. No attempt is made here to describe in detail the molecular orbitals or metal symmetry involved in the bonding of CO to the metal. Vibrational frequencies have been identified for the linear and bridged structures in metal carbonyl complexes as 2200-2000 cm^{-1} and 1880-1820 cm^{-1}, respectively.[20] Blyholder[19] suggests it is not necessary to postulate the existence of a bridge structure to explain the lower

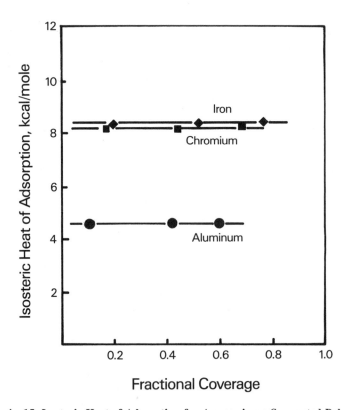

Figure 4 - 15 Isosteric Heat of Adsorption for Ammonia on Supported Polynuclear Metal Complexes

Chapter 4 Classical Methods

frequency relaxations observed for CO on metals. He speculates a single molecular orbital model for the bonding of CO may account for the different relaxation frequencies. The higher frequencies represent CO bonded to the middle of a lattice plane; whereas the lower frequency bands represent CO bonded to edges, corners, and defects.

The numerical results for a nine atom array of Ni metal atoms do model higher frequency relaxations by the CO bonding to the central Ni site or the lower frequency site for the CO bonding to the one of the other eight "edge" atoms. Thus qualitatively the molecular orbital model does "produce" the expected behavior in terms of the relative positions of the vibrational bands. Additionally, the model does explain (1) the spectral differences between CO adsorbed upon evaporated and supported metals; (2) the change in IR band positions as a function of coverage; (3) the effects of adsorbing other gases in addition to previously sorbed CO; and (4) the band shifts observed between pure metal and alloys.

To gain a deeper understanding of the model, we must review the molecular orbital picture of CO bonded linearly to a transition metal atom (see Fig. 4 - 16). The bonding orbitals between the carbon and the oxygen are the 1σ resulting from a carbon sp hybrid and an oxygen p and the 1π from the carbon p_x and p_y and oxygen p_x and p_y. The oxygen shows a 2σ lone pair orbital and occupied 2p orbitals; whereas the carbon atom has a full 5σ orbital and empty $2\pi^*$ antibonding orbitals. The full 5σ orbital of the carbon may donate a dative electron pair to an empty d orbital of the metal atom. The formation of this dative σ bond between the carbon and the metal atom puts a formal negative charge on the metal and this charge is dispersed via back-donation through the metal d orbitals into the CO antibonding orbitals. These π^* antibonding orbitals do show energies less than those of the originating metal d orbitals. This back-donation increases the strength of the metal carbon bond at the expense of decreased strength in the carbon-oxygen bond. It has been observed in the IR spectra of metal carbonyl compounds that for increased back-donation to the π^* antibonding orbitals, the vibrational frequency of the CO bond decreases, thus supporting the argument.

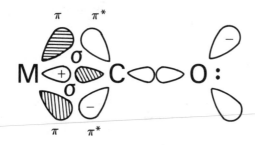

Figure 4 - 16 Molecular Orbital Diagram of CO

To apply this molecular orbital information of CO to the process of chemisorption, a metal atom on the surface is pictured as the central metal atom in a complex with the surrounding metal atoms and the chemisorbed CO as the ligands. The "ligand" metal atoms surrounding the central metal atom will have partially filled d orbitals available for back-donation into the π^* orbitals of the CO. The extent of the competition for available electrons between the surrounding metal atoms to enter the π^* orbitals will determine the frequency and extinction coefficient of the chemisorbed species. Thus, sorbate species occupying different positions on the crystal faces will experience varying degrees of electron competition among the available metal atoms.

It is useful to summarize the observations regarding the behavior of CO on metals:

1. C-O stretching frequency shifts to higher frequencies as the fractional coverage increases.
2. The relative intensities of bands for CO on evaporated and supported films are different.
3. Adsorption of gases *after* preadsorbed CO shifts the positions of the CO IR absorption bands.

Let us consider now the spectrum of CO on crystals on Ni. On the $\{111\}$ plane or $\{100\}$ plane of Ni there will be nine or eight adjacent Ni atoms, respectively. With these large numbers of metal atoms competing for charge to go into the π^* antibonding system of the sorbed species, only a small amount of the charge actually reaches the CO molecule; thus the CO stretching frequency is near 2000 cm^{-1}. Conversely, at an edge, corner, or defect site, the decreased competition among metal atoms to disperse the charge results in more of the electrons to be donated back into the sorbate system, thus decreasing its IR frequency. As a result of increasing charge density in the π^* system, the extinction coefficient is also speculated to increase. Thus one expects to observe increasing extinction coefficients for the decreasing frequency of the CO vibrations.

The sorption of CO onto silica supported Ni and evaporated metallic Ni films can show quite different IR spectra depending upon the treatment of the samples. Consider a supported Ni catalysts for which the reduction conditions ensure sintering and recrystallization of the Ni; the theory predicts a large, higher frequency band and a small, lower frequency band. These results have been reported by Eischens[21] for Ni on silica. Consider now an evaporated Ni film quenched after deposition. This sample is expected to produce an amorphous material showing many defects. These defects should give rise to the lower frequency band. Again, Eischens[21] reports the same.

The position of the IR bands with surface coverage is considered next. Two explanations are possible to describe this behavior. First, the sorbed molecules on the surface may compete for the electrons of the surface atoms which may be back-donated. As the surface coverage increases, the degree of competition for these electrons also increases; thus the observed frequency increases with coverage. Secondly, those sites which give rise to the lowest frequency are the most energetic sites and will be covered *first*. Both kinds of behavior are probably observed.

Next, we consider the effects of other gases upon the vibrational spectrum of sorbed CO. Here we must characterize the coadsorption process as requiring electrons or donating electrons. If the coadsorbate requires electrons, the theory predicts the CO frequency should increase with extent of coadsorption and conversely if the coadsorbate donates electrons. Eischens[21] reports the coadsorption of O_2 to form oxide ions, a process requiring electrons to shift the CO bands to higher frequency, whereas the addition of H_2, a process which donates electrons, to have the reverse effect on the CO spectrum.

Finally, we discuss CO sorption on mixed metal clusters (alloys). Consider Eischens' data on the Cu-Ni system. The electronic picture of copper shows its d orbitals to be filled; whereas the Ni system has only partially filled d orbitals. It is expected the copper would donate electrons to the π^* system, thus decreasing the observed frequency. Eischens' data[22] shows a 90%Ni-10%Cu alloy to have a spectrum shifted 40 cm^{-1} to the lower frequencies than that observed for pure Ni.

The effect of charge dispersed by the surface metal atoms was investigated further by Blyholder. He measured the vibrational spectrum for CO chemisorbed on metal films of the first row of the transition metals (V, Cr, Mn, Fe, Co, Ni, Cu). The observed infrared band positions are reported in the Table 4 - 4. For the first-row transition elements, the energy level for the d orbitals decrease with increasing atomic number. This trend in d orbital energy is confirmed from a consideration of the valence state ionization potential (VSIP) which is a measure of the energy necessary for the removal of a d electron from the 3d orbital of the metals. Increasing values of the VSIP imply more energy is *required* to remove the 3d electron; thus that electron was of lower energy. Since the motivation for back-donation of the metal 3d electron into the π^* system was a favorable energy *decrease*, this potential for back-donation lowers as the energy level of the *donating* system becomes less.

Table 4 - 4
IR Spectra of CO Sorbed to Transition Metals in First Row

Metal	VSIP (cm^{-1})	High Frequency Band (cm^{-1})	Low Frequency Band (cm^{-1})
V	51,400	1940	1890
Cr	57,900	1940	1880
Mn	64,100	1950	1890
Fe	70,000	1980	1900
Co	75,000	2000	1880
Ni	80,900	2075	1935
Cu	86,000	2120	----

Reprinted with permission from G. Blyholder and M. C. Allen, *J. Am. Chem. Soc.*, **91**:12 (1969), p. 3158. Copyright 1969, American Chemical Society.

Molecular orbital calculations by Blyholder[19] attempted to model the bonding of carbon monoxide to a nine-atom array of metal atoms showing the characteristics of nickel crystal and a chromium crystal. The goals of these calculations were to estimate the C-O bond strengths for sorption to a metal "middle site" and to a metal corner site. These orbital calculations would be compared to the IR spectra of CO chemisorbed to the first-row transition metals so as to explain the observed changes in the frequencies of the two bands. A useful quantity to interpret the molecular orbital calculations in the bond order which is a measure of the amount of multiple bond character in a bond. These bond orders, defined as follows,

$$P = \sum_i N_i C_i^* C_i$$

where

N_i = number of electrons in the *ith* molecular orbital
C_i^*, C_i = coefficients of atomic orbitals ψ^*, ψ in the *ith* molecular orbital

have been correlated with bond lengths and *vibrational* force constants.

Blyholder's results, given in Table 4 - 5, show the bond order for CO sorbed to a middle site (i.e., a high frequency relaxation 2000 cm^{-1}) increases with increasing VSIP in cm^{-1} (ergo atomic number); whereas the bond order for CO sorbed to a corner site increases only slightly for increasing VSIP. Recognizing that increased bond orders imply stronger force constants and higher frequency, these results of bond orders show the higher frequency band should shift to higher frequencies with increasing VSIP, whereas the lower frequency band should shift upwards only slightly. The data of Table 4 - 5 are in qualitative agreement with these MO predictions.

Table 4 - 5
Summary of MO Calculation on 9-atom Crystals

Crystal	Bond Orders for CO	
	Middle Site (high frequency)	Corner site
"Ni" VSIP= 0.41 au	0.304	0.276
"Cr" VSIP= 0.30 au	0.268	0.264

Reprinted with permission from G. Blyholder and M. C. Allen, *J. Am. Chem. Soc.*, **91**:12, (1969), p. 3158. Copyright 1969, American Chemical Society.

Temperature Programmed Desorption

Another tool wellsuited to describe the energy of sorption is temperature programmed desorption (TPD). The procedure is to desorb a selective adsorbate from a catalyst sample and note the rates at which the desorption occurs as a function of temperature. One apparatus especially useful is the thermal gravimetric apparatus. A high flow of inert gas is established over the shallow bed of sample and the temperature is increased at a fixed rate, say β° K/minute. The amount of gas deserbed is recorded versus the sample temperature. This procedure is repeated for several different programming rates to determine the temperature at which maximum *rate* of gas desorbs. It is desired to calculate the activation energy for *desorption* so as to characterize the strength of the sorbent-surface bond.

The assumptions for this development are as follows:

1. *Homogeneous* surface for adsorption, that is,

$$k_d = k_{do}\ exp[-E_d/RT]$$

is not a function of coverage (θ).

2. Readsorption of the desorbed gas does not occur; that is, the gas flowrate through the bed is high.
3. Concentration of the adsorbate in the gas phase is uniform through the bed.
4. Desorption rate is first order in coverage.
5. Linear temperature increases with time:

$$T = T_o + \beta t$$

The appropriate material balance on the desorbing species is

$$-V_m\ d\theta/dt = k_{do}\ exp[-E_d/RT]\theta \qquad (4 - 71)$$

but

$$d\theta/dt = (dT/dt)\ d\theta/dT = (\beta)\ d\theta/dT \qquad (4 - 72)$$

so now

$$d\theta/dT = -k_{do}\theta exp[-E_d/RT]/(\ \beta V_m) \qquad (4 - 73)$$

At the maximum in desorption rate, $d^2\theta/dT^2 = 0$, so

$$[d^2\theta/dT^2]\big|_{Tm} = -\{k_{do}/\beta V_m\}\{d\theta/dT + E_d\theta/RT^2\}\ exp[-E_d/RT]\big|_{Tm} = 0 \qquad (4 - 74)$$

Rearranging we have

$$(d\theta/dT)\big|_{Tm} = -E_d\theta/(RT_m^2) \qquad (4 - 75)$$

but $d\theta/dT$ is given by

$$(d\theta/dT)\big|_{Tm} = (k_{do}\theta/\beta V_m)\ exp[-E_d/RT] = -E_d\theta/RT_m^2 \qquad (4 - 76)$$

$$-E_d/RT_m = ln(\beta/T_m^2) + ln[E_dV_m/Rk_{do}] \qquad (4 - 77)$$

which may be rearranged as follows:

$$2lnT_m - ln\beta = E_d/RT_m + ln[E_dV_m/Rk_{do}] \qquad (4 - 78)$$

Eq. (4 - 78) suggests the data for which the maximum rate of desorption occurs at temperature T_m for each programming rate β may be plotted as $(2lnT_m - ln\beta)$ on the ordinate versus $(1/T_m)$ on the abscissa. The slope will be E_d/R ; whereas the intercept is $ln(E_dV_m/Rk_{do})$. Thus, one set of data yields E_d and V_m/k_{do}.

Consider the temperature programmed desorption of ethylene from alumina.[23] The amount of ethylene sorbed onto the "clean" alumina was described by a conventional volumetric technique after which the catalyst was evacuated at prescribed conditions. Next, a set flow of He was directed through the catalyst bed into a TC detector with the effluent sent to a liquid nitrogen trap. The temperature of the catalyst was programmed at constant rates, beta, during which the ethylene was desorbed. Since the flowrate of He was constant during the desorption, the response of the TC cell is proportional to the desorption rate of ethylene from the alumina. During the typical desorption run, the rate would attain a maximum at a temperature, T_m, for each type of sorption site (see Fig. 4 - 17). These data are summarized in Table 4 - 6 for the lower temperature peak.

Notice in Figure 4 - 17 two peaks are apparent. The shapes of the curves change with programming rates as predicted by Eq. (4 - 78). The relative strengths of the adsorption sites may be inferred from the location of the desorption rate maxima, T_m. Peaks occurring at high values of T_m infer these species are held tightly to the surface. Added insight to the relative site strengths comes from an evacuation at a temperature higher than the lower temperature peak prior to the TPD. The resulting TPD spectrum "fits" nicely into the higher temperature peak of the original spectrum.

Quantitative estimates of the site energies are possible using rather simple theory. From Eq. (4 - 78) the temperature of the desorption rate maximum and the rate of progamming are plotted in Figure 4 - 18 to determine the energy of desorption, E_d, and the rate constant preexponential factor for desorption divided by the volume of the monolayer, k_{do}/V_m. For the first peak these values are 26.8 kcal/mole and 1.6 x 10^{-15} sec^{-1} and for the second peak the energy is 36.4 kcal/mole with the same value for k_{do}/V_m. The olefin occupies 2.8% of the total surface with 60% of these sites corresponding to the lower energy site. These quantitative E_d data agree with the qualitative results.

Recently, several attempts at modeling the TPD process within a porous catalyst have revealed serious deficiencies in the procedure.[24, 25] Gorte described the values of dimensionless groups for which the mass transport effects modeled by the group could be neglected. His analysis showed that the TPD spectra from some porous catalysts were tainted by the effects of adsorbate readsorption within the pores to produce activation energies for desorption which were low. We[26] show evidence to support Gorte's claims in three samples of silica-alumina (20 wt% Al_2O_3) having average pore sizes of less than 30 angstroms, A; 50 angstroms, B; and 60 angstroms, C. The E_d for ammonia TPD from these samples were as follows:

A: 9.2 ± 0.6 kcal/mole
B: 13.5 ± 1.3 kcal/mole
C: 14.3 ± 1.3 kcal/mole

These results suggest the sample showing qhe highest probability of readsorption in the pores, sample A, also showed the smallest E_d. These three samples were prepared by steaming a fresh catalyst, sample A, at two different severities of steam treatment to produce samples B and C; the chemical constitution of the samples were the same. It is clear from these results that the effects of pore size, hence adsorbate readsorption, may cloud the results of the TPD experiment.

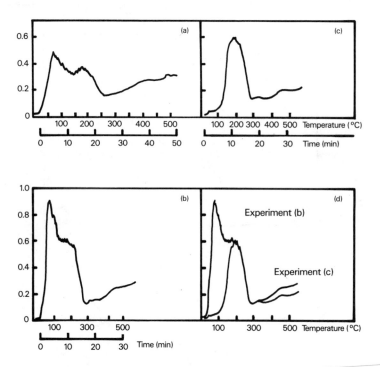

Figure 4 - 17 Temperature Programmed Desorption of Ethylene from Alumina
Reprinted with permission from Y. Amenomiya and R. J. Cvetanovic, *J. Phys. Chem.*, **67**, (1963), p. 144. Copyright 1963, American Chemical Society.

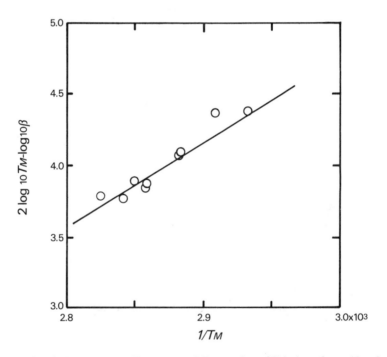

Table 4 - 6 Temperature Programmed Desorption of Ethylene from Alumina

Temperature ($^\circ$C)	Pressure (Torr)	β ($^\circ$C/min)	T_M($^\circ$C)
24.9	6.13	10.08	74
24.8	6.34	20.70	79
24.7	6.37	5.05	68
25.2	5.28	5.14	72
26.2	5.63	21.05	81
26.5	5.38	10.40	74
25.8	5.96	16.03	77
24.2	9.00	16.05	78

Reprinted with permission from Y. Amenomiya and R. J. Cvetanovic, *J. Phys. Chem.*, **67**, (1963), p. 144. Copyright 1963, American Chemical Society.

Figure 4 - 18 Temperature Programmed Desorption of Ethylene from Alumina

Reprinted with permission from Y. Amenomiya and R. J. Cvetanovic, *J. Phys. Chem.*, **67**, (1963), p. 144. Copyright 1963, American Chemical Society.

Chapter 4 Classical Methods

Conclusions

Classical techniques have been described to characterize the surfaces of catalysts. Volumetric apparatus are useful to measuring the specific total and active surface areas of adsorbents. Alternatively, gravimetric devices may be used to measure the adsorption amounts. For porous supports, the pore size distribution may be characterized by two complementary techniques: nitrogen desorption and mercury intrusion. Energies of sorption may be inferred by vibrational spectroscopy, equilibrium adsorption at several different temperatures, and by thermal desorption spectra. These surface characterizations may be used to illuminate activity data of fresh and aged catalyst.

REFERENCES

1. Brunauer, S., L. S. Deming, W. E. Deming, and E. Teller, *J. Am. Chem. Soc.*, **52**, (1940), p. 1723.
2. Poehlein, S. R., M. S. thesis, Georgia Institute of Technology, Atlanta, Georgia (1984); Stull, J. O. and M. G. White, *O. E. D.*, Vol. 11, (1986), p. 325.
3. Brunauer, S., P. Emmett, and E. Teller, *J. Am. Chem. Soc.*, **60**, (1938), p. 309.
4a. Barrett, E. P., L. G. Joyner, and P. P. Halenda, *J. Am. Chem. Soc.*, **73**, (1951), p. 373.
4b. Faas, G. S., M. S. thesis, Georgia Institute of Technology, Atlanta, GA, (1981).
4c. O'Neil, C. S., *American Laboratory*, June, (1985), p. 41.
4d. Mingle, J. O. and J. M. Smith, *Chem. Eng. Sci.*, **16**, (1961), p. 31.
5. Wheeler, A., *Adv. Catal.*, **2**, (1951).
6. Hightower, J., "The Uses of Heterogeneous Catalysis, Short Course," Houston, Texas (1975).
7. Ritter, H. L. and L. C. Drake, *Ind. Eng. Chem. Anal. Ed.*, **17**, (1945), p. 787.
8. Barrer, R. M., N. McKenzie, and J. S. Reay, *J. Colloid Sci.*, **11**, (1956), p. 479.
9. Spenadel, L. and M. Boudart, *J. Phy. Chem.*, **64**, (1960), p. 205.
10. Mears, D. E. and R. C. Hanford, *J. Catal.*, **9**, (1971), p. 125.
11. Benesi, H. A. and B. H. C. Winquist, *Adv. Catal.*, **27**, (1978), p. 97.
12. Rosenthal, D. J., M. S. thesis, Georgia Institute of Technology, (1985); with M. G. White and G. D. Parks, *Am. Inst. Chem. Eng. J.*, **33**:2, (1987), p. 336.
13. Mapes, J. E. and R. P. Eischens, *J. Phys. Chem.*, **58**, (1966), p. 1059.
14. Rooney, J. J. and R. C. Pink, *Trans. Farad. Soc.*, **58**, (1962), p. 1632.
15. Wilhoit, R. C., *J. Chem. Educ.*, **44**, (1967), pp. A571, A629, A685, A853.
16. Armstrong, G. T., *J. Chem. Educ.*, **41**, (1964), p. 297.
17. Gravelle, P. C., *Adv. Catal.*, **22**, (1972), p. 191.

18. Beckler, R. K., Ph. D. thesis, Georgia Institute of Technology, (1987).
19. Blyholder, G. and M. C. Allen, *J. Am. Chem. Soc.*, **91**:12, (1969), p. 3158; Blyholder, G., *J. Phys. Chem.*, **68**, (1964), p. 2772.
20. Kavtaradze, N. N. and N. P. Sokolova, *Zh. Fiz. Khim.*, **40**, (1966), p. 2957.
21. Eischens, R. P., S. A. Francis, and W. A. Pliskin, *J. Phys. Chem.*, **60**, (1956), p. 194.
22. Eischens, R. P., *Z. Elektrochem.*, **60**, (1956), p. 782.
23. Amenomiya, Y. and R. J. Cvetanovic, *J. Phys. Chem.*, **67**, (1963), p. 144.
24. Gorte, R. J., *J. Catal.*, **75**, (1982), p. 164.
25. Demmin, R. A. and R. J. Gorte, *J. Catal.*, **90**, (1984), p. 32.
26. Brinen, J. L., M. S. thesis, Georgia Institute of Technology, (1985); with M. G. White and A. M. Schaffer, ACS National Meeting, Symposium on Role of Trace Elements in Hydrocarbon Processing, Paper 71 (April, 1986).

PROBLEMS

1. Show that if p is much less than p_o, then the BET isotherm reduces to the Langmuir isotherm.

2. Oxygen is chemisorbed on a nonporous charcoal sample showing the structure of graphite under the conditions of $150°K$ for five hours. This sample is heated to $500°K$ which causes all the chemisorbed oxygen to desorb in the form of carbon dioxide. What is the minimum active surface area of this sample assuming the charcoal is in the form of one million microspheres and if the weight loss during desorption of the carbon dioxide is 11 micrograms? Find the average radius of the spheres.

3. Assuming that a suppported Pt catalyst shows octahedral crystallites for which only the {*111*} face is exposed to the gas phase, calculate the dispersion as a function of crystallite size.

4. Rhodium catalysts may chemisorb CO in either the bridge structure (I) or the linear structure (II). The hydrogenation of CO has been postulated to proceed as follows:

$$CO + nH_2 \Leftrightarrow CH_3OH \quad \textit{(Type I)}$$
$$CO + nH_2 \Leftrightarrow CH_4 + H_2O \quad \textit{(Type II)}$$

How will the methanol/methane selectivity depend upon the particle size for a supported Rh catalysts according to the theory advanced by Blyholder for CO sorption to edge and face sites?

5. An adsorption apparatus (Fig. 4 - 1) consists of four volumetric bulbs, a sample holder, and interconnecting tubing. A catalyst is placed in the sample holder and isolated form the system. Nitrogen is added until the pressure is 560 torr. The valve to the sample holder is opened and the pressure equilibrates at 350 torr. A

second experiment with the same initial pressure shows a final pressure of 450 torr. How many moles are added in each case and what is the volume of the sample holder? The volume of the four bulbs is 120 cc the volume of the tubing is 4.4 cc and the temperature of the bulbs is $25°C$. In the first run the sample temperature is $77°K$ and the second shows a temperature of $25°C$.

6. The mercury penetration data given in Table 4 - 7 were obtained for a 0.624 g sample of uranium dioxide formed by sintering particles at $1000°C$ for 2 hours. Since the particles were nonporous, the void space was the macropores. At the beginning of the experiment (p = 1.77 psia) the amount of mercury displaced by the sample was 0.19 cc. Calculate the porosity and pore volume distribution of the pellet. Used with permission from J. M. Smith, *Chemical Engineering Kinetics*, 3rd Edition (New York: McGraw-Hill Book Co., 1981), p. 358.

7. X-ray diffractions with radiation having a wavelength of 3 angstroms will diffract what angles from the {100} plane of a simple cubic lattice showing a lattice constant of 5.0 angstroms?

8. Given the following data for mercury intrusion into a sample, calculate the pore radius distribution and the most probable pore radius.

Table 4 - 7
Mercury Intrusion Data for Problem 8

Pressure (psia)	Intrusion (cc/g)
2,000	0.010
5,000	0.012
10,000	0.014
15,000	0.017
20,000	0.020
25,000	0.024
30,000	0.029
35,000	0.034
40,000	0.039
44,000	0.051
48,000	0.070
51,000	0.125
53,000	0.150
55,000	0.191
57,000	0.265
59,000	0.335
61,000	0.390
63,000	0.405
65,000	0.410

Used with permission from J. M. Smith, *Chemical Engineering Kinetics*, 3rd Edition (New York: McGraw-Hill Book Co., 1981), p. 358.

9. The nitrogen physisorption data at $77°K$ on 1 gram of silica-alumina is reported in the following table. Calculate the specific surface area of this sample.

p/p_o	V_{ads} scc
0.05	24.25
0.10	25.60
0.15	27.11
0.20	28.80
0.25	30.58
0.30	32.90
0.35	35.44

10. A series of supported metal catalysts are exposed to hydrogen at increasing partial pressures to define adsorption isotherms. The monolayer adsorptions are given in the following table. For these catalysts calculate the metal dispersion and the average crystallite size assuming the cube model. Be specific in detailing all the assumptions necessary to answer these questions. Also give the metal surface area/gram of metal. In determining the average area/atom, find the arithmetic average of the {100}, {110}, and {111} planes. You may find the following information useful.

Catalysts Loading	V_m, scc/g
5% Ru	1.5
5% Rh	3.0
10% Ir	4.5
10% Rh	5.3

Metal	Lattice	Metal Atom Radius, A	Density of Metal, g/cc
Ru	HCP	1.32	11.9
Rh	FCC	1.34	12.44
Ir	FCC	1.35	22.5

11. The CO infrared spectra of a promoted and unpromoted copper oxide are shown in Figure 4 - 19. It was found that the support and the SnO promoter do not chemisorb CO. The dotted line in Figure 4 - 19 shows the resolved peak from the subtraction of the CuO_x/SiO_2 peak form the parent asymmetric peak. Notice that CuO_x/SiO_2 does not show a peak near 2073 cm^{-1} as does the tin promoted copper oxide catalyst. Figure 4 - 20 shows the IR spectrum of chemisorbed CO onto the promoted copper catalyst which has been subjected to mild oxidizing conditions at 298 K and a partial pressure of oxygen equal to 52.5 millitorr. The spectra of the chemisorbed CO were recorded at 0, 48, 96, and 120 hours. The conditions for the CO sorption were 45 torr at room temperature. The low frequency peak at 2073 cm^{-1} disappears within the first 96 hours and the higher frequency peak shift from 2137 to 2127 cm^{-1} and decreased in size.

Identify the two CO peaks in the promoted copper catalysts, if one of the peaks is a combination of two peaks, indicate as such. Using the molecular orbital theory advanced by Blyholder, explain the strengths of the metal-carbon bonds as inferred from the CO relaxations in the promoted and unpromoted catalysts. Which of the sites would favor overoxidation of the CO to CO_2 based on this MO theory?

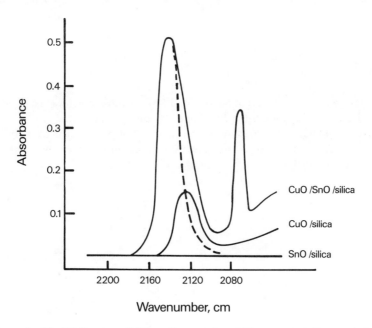

Figure 4 - 19   IR Spectra of CO on Promoted and Unpromoted Copper Oxide

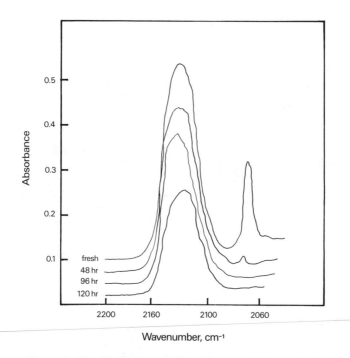

Figure 4 - 20   IR Spectra of CO on Promoted Copper Oxide

Explain the results of the slow oxidation tests. Note: These catalysts were activated by the following procedure:

Oxygen for 30 minutes at 50 Torr and $325°C$ followed by evacuation at the same temperature for 30 minutes. Propylene was introduced at 50 Torr and $325°C$ for 30 minutes, then the catalysts were evacuated at the same temperature for 30 minutes. After evacuation the catalysts were ready for analysis.

CHAPTER 5

SURFACE SCIENCE METHODS FOR CHARACTERIZATION

Surface science instruments have become an important part of the arsenal of techniques for the heterogeneous catalysis scientist. Certainly, an understanding of the surface structure is important to the proper interpretation of catalytic rate data. The morphology of the clean surface may be described by a combination of tools. We include only four of the many available: low energy electron diffraction (LEED), Auger electron spectroscopy (AES), X-ray photoelectron spectroscopy (XPS), and extended X-ray absorption fine structure (EXAFS). Each will be described in the following sections.

Electron Diffraction from Surfaces (LEED)

The study of "clean" single crystals has often led to a better understanding of the catalysis of the polycrystalline solid. Kemball and Bond[1] show a measure of consistency between the catalytic rates on evaporated metal films and polycrystalline solids. The motivation for the study of such simple single crystal systems is the ease with which they may be characterized relative to the polycrystalline samples. The technique of low energy electron diffraction has been used to probe the surface morphology of single crystals and the chemisorbed layers to the same.

Single Crystal Approach
The approach of LEED is to employ single crystals of the catalyst which have been cut to expose the desired crystal face. This face is polished and cleaned of the surface contaminants and then exposed to the appropriate conditions. This analysis requires conditions which are removed from the operating conditions of the commercial catalysts. For example, the commerical catalysts are polycrystalline solids, supported on high surface area polycrystalline oxides which show vastly different surface morphologies from the single crystal samples analyzed in the LEED apparatus. The pressures in the industrial reactor are a factor of 10 billion times those established in the LEED apparatus. However, the value of the LEED experiment and other ultrahigh vacuum studies is the knowledge gained regarding the fundamental relationship between the surface microstructure and the catalysis. The controlled, albeit unrealistic, conditions of the ultrahigh vacuum environment enable

the surface chemist to ferret out the effects upon the catalysis of surface morphology from the melange of other effects which are present at conditions typical of the industrial reactor.

The present summary focuses on the results obtained for single crystals of metals. Metals show unusually high "efficiency" of catalytic reaction for the available sites; for example, almost every surface metal atom is active for some simple reactions such as deuterium exchange. The same statement is not true for oxides where less than 0.1% of the available surface may be active. Moreover, metals contain only one type of atom; whereas oxides contain at the very least two types of atoms, plus they may contain surface hydroxyls. The reader is referred to other reviews and texts for in-depth discussions of the principles of LEED and applications.[2a, 2b, 2c]

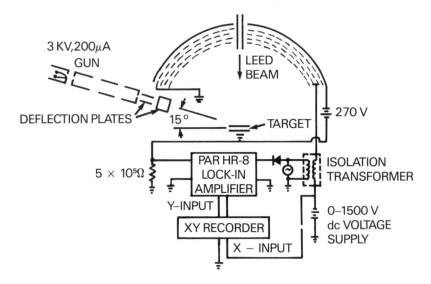

Figure 5 - 1 LEED Apparatus

Used with permission from P. W. Palmberg, *Appl. Phys. Lett.*, **13**, (1968), p. 183.

Theory of Operation and Applications

For LEED, electrons are accelerated, initially, by low potentials (50 to 100 V) to strike a single crystal target (see Fig. 5 - 1). Because these electrons are of such low energy, the majority of these particles do not penetrate the bulk but rebound from the surface in the form of a diffraction pattern. The electrons are sufficiently "small" so as to probe the atomic texture of the surface resulting in a description of the atomic arrangement of the *surface* atoms and the lattice spacing between the atoms.

Typical LEED data are electron intensity patterns developed in space. These patterns may be detected by a fluorescent screen (Fig. 5 - 1) or counted over a set time period in a Faraday cup oriented at a selected position in space. If a screen is used, photographs of the screen are recorded and the positions of the dot pattern are noted. Figure 5 - 2 shows the patterns developed by normal incidence from a tungsten surface represented by the ball models. Note for the same surface, W {110}, increasing the acceleration voltage of the incident electron beam from 75 V to 300 V causes more of the dot pattern to be visualized. The patterns can be predicted for simple surfaces using the Bragg Law for diffraction applied to two-dimensional surfaces. The details are given by May in Reference 2a.

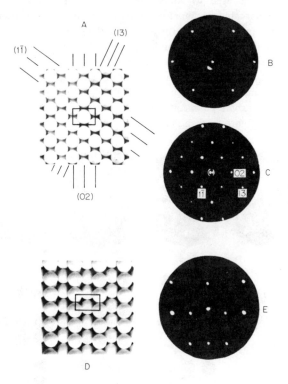

Figure 5 - 2 LEED Patterns for Clean Tungsten Crystal

Low energy electron diffraction patterns at normal incidence. (a) Ball model of W(110) face. Some of the net lines (hk) are indexed in terms of a centered rectangular unit mesh (outlined). (b) Clean W(110), 75 V. Diffuse brightness and central bright spot are caused by light from electron gun filament. (c) Clean W(110), 300V. (d) Ball model of (112) surface, the third densest of the bcc lattice. (e) Clean W(112) at 90 V. Note asymmetric intensities of the hk and hk beams. There is strong scattering contribution from the exposed second layer which is positioned asymmetrically. Used with permission from J. W. May, *Adv. Catal.*, **21**, (1970), p. 150.

In one application of this technique, the morphology of the "clean" surface has been examined then exposed to a chemisorbing gas for which this monolayer was probed by LEED to define the positions of the overlayer with respect to the atoms of the support. The results of these studies have led to a better understanding of adsorption on solids. Figure 5 - 3 shows the LEED patterns for Ni {110} exposed to O_2 at room temperature. The pattern changes drastically for exposures of the oxygen at different coverages. From these patterns the placement of the oxygen atoms may be deduced easily in relation to the underlying nickel atoms. If the oxygen exposure is continued, an oxide overlayer develops which perturbs the nickel ion overlayer from the lattice points of the metal (Fig. 5 - 4).

LEED has been used to study the surface morphology as a function of coverage. Most remarkable is the discontinuous manner by which surface coverage may change with increasing pressure of adsorbate as in hydrogen on nickel {110}. One explanation for this discontinuity in surface coverage is a change in the morphology of the overlayer with respect to the bulk. That is, at low hydrogen partial pressures the H atoms may assume positions above the surface requiring a greater number of metal atoms per H atom than at higher partial pressures. The resulting change in sorbate atom/sorbent atom stoichiometry causes a change in the observed coverage. Figure 5 - 5 shows this transition for CO sorbed on Pd {100}.

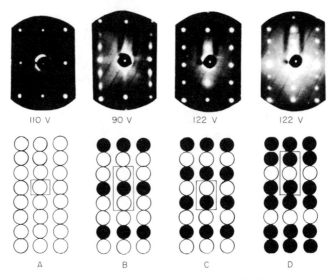

Figure 5 - 3 LEED Patterns for Oxygen on Ni (110) Crystal

The (110) Ni surface exposed to oxygen at room temperature. Sequence a through d shows LEED patterns at normal incidence as oxygen is adsorbed and as coverage increases. Starting with a clean surface, (a), the pattern changes to (3x1), (b), after oxygen exposure 0.5 L (0.5 x 10^{-6} Torr-sec). Further exposure gives a (2x1) pattern completely developed after 0.8 L, which changes to a second (3x1) at 5 L (c and d). Used with permission from J. W. May, *Adv. Catal.*, **21**, (1970), p. 150.

Chapter 5 Characterization by Surface Science Techniques

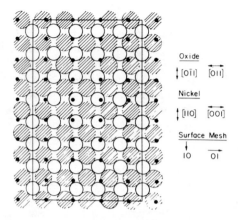

Figure 5 - 4 LEED Data for Oxide Overlayer on Nickel

Sketch of two-dimensional pseudooxide which forms on Ni(110) at room temperature. Centers of substrate nickel atoms are shown as black dots, overlayer Ni atoms as open circles, and O atoms in the overlayer as hatched circles. The combined oxide monolayer and substrate make up a (9x4) mesh contained in the rectangle having dimensions of $9a_o/(2)^{1/2} = 22.4$ ang. by $4a_o = 14.1$ ang. Used with permission from J. W. May, *Adv. Catal.*, **21**, (1970), p. 150.

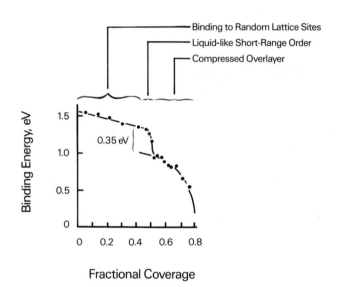

Fractional Coverage

Figure 5 - 5 Phase Transition of Chemisorbed CO on Pd (100)

A discovery using LEED is that two-dimensional phase transitions occur in adsorbate layers as the interactional forces change with coverage. Here, at about one-half monolayer coverage on Pd(100), CO switches from a structure with bonding to definite lattice sites to a compressed overlayer out of registry with the substrate.

Another remarkable observation arising from LEED studies is the effect of some chemisorbing gases to cause surface reconstruction.[3] Consider sorption of hydrogen onto the Pt [6(111)x(100)] face where LEED shows the rearrangement of the surface Pt atoms. This reconstruction of Pt occurs at a characteristic temperature and pressure for which it has been suggested to describe a surface phase-transition phenomenon. For the case of hydrogen chemisorption, the reconstructed layer shows a repeat distance of twice that of the original layer before reconstruction. Other examples of surface dissolution/reconstruction include the solvation of oxygen atoms into a nickel crystal followed by formation of an oxide overlayer.[4]

Electron Spectroscopy of Solids (AES/XPS)

Electron spectroscopy, as the term is used to describe the techniques to characterize catalysts, has enjoyed great success for describing the surface region of heterogeneous catalysts. Two useful tools (AES/XPS) offer the unique ability to probe to a depth of no more of 50 to 100 angstroms from the surface and thus offer a signficant advantage over bulk techniques, such as X-ray diffraction, for relating catalyst characterization to observed catalytic properties.[5a, 5b, 6] In this section, we show the atomic processes attending the two techniques, the apparatus for measuring the same, and some practical examples in the literature showing the application.

Atomic Processes - XPS

Consider first the atomic processes related to XPS. The incident radiation is X-ray which causes a core level electron, say a K level, to be excited from the atom. The kinetic energy of these ejected electrons are measured in the appropriate electron energy analyzer. The energy of the excited electron depends upon the atomic environment, which we seek to describe, as well as the work function of the spectrometer and the energy of the incident radiation. In most cases the work function of the spectrometer can be held constant and calibrated against standards; whereas the energy of the radiation is known once the wavelength is measured. Thus, the kinetic energy of the ejected electrons may be related directly to the nature of the atomic environment. Figure 5 - 6 (panel 1) describes the XPS process in terms of electron energy levels of the host atom. These levels are determined largely by the identity of the host atom (i.e., nuclear charge) and the oxidation state. Secondary effects such as the charge/identity of the neighboring species will cause small shifts in the energy of the ejected electron. The excited electron will assume a trajectory which will encounter other atoms/ions. With each encounter the electron faces the possibility of capture by the atom/ion. Clearly, only those host atoms/ions near the surface will produce excited electrons that will not be captured. In reality some encounters of excited electron with atoms do not result in capture; thus the escape depth is several atomic layers, depending upon the material. Figure 5 - 7 shows the XPS lines for iodine in a sample of iodostearic acid.[6] This material forms double layers of controlled thickness, and by suitable treatments, one to ten double layers

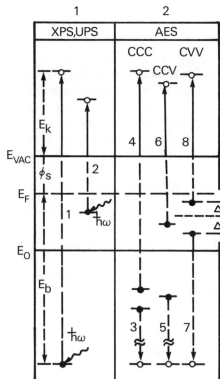

Figure 5 - 6 Energy Level Diagram for XPS and AES Atomic Processes

Used with permission from *Experimental Methods in Catalytic Research*, Vol. III., Ed. by R. B. Anderson and P. T. Dawson (New York: Academic Press, Inc., 1976), p. 50.

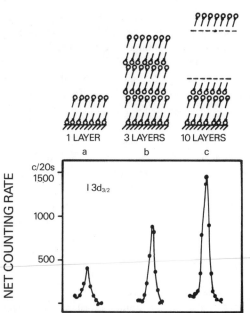

Figure 5 - 7 Mean Escape Depth in Double Layers of Iodostearic Acid

Used with permission from K. Sieghbahn et al., *ESCA: Atomic, Molecular, Solid State Structure Studied by Means of Electron Spectroscopy* (Uppsala: Almqvist & Wiksells, 1967), p. 140.

could be developed in the samples to be examined by XPS. The data showed that the iodine XPS peak intensity increased with an increasing number of double layers. However, as the number of double layers approached ten, the XPS peak intensities remained almost constant. The escape depth for this sample was no more than ten double layers. Clearly, other nuclei will show different escape depths depending upon the energy of the excited electron and the "cross sectional" area of the atoms for capture of the electron. Thus, XPS is a tool for characterizing the surface "region."

Atomic Processes - AES

The atomic process for AES is similar to that for XPS, save for the exciting radiation may be a beam of electrons or other radiation which causes a core level electron to be excited to a high state within the host atom. This vacancy caused by the excited electron is subsequently filled by an electron from the outer shell relaxing to a lower energy level thus causing another outer shell electron to leave the host atom. These electrons may leave the host atom as the Auger electron whose kinetic energy is measured by an electron energy analyzer. AES involves two-electron transitions; thus several cases may be generated to describe the overall process which yields no "net" radiation from the host atom. Refer to Figure 5 - 6, panel 2, for an illustration of three AES processes represented by the paired transitions (3:4), (5:6), and (7:8). In these transitions, the core hole may be filled by a "downward moving" core electron (3 and 5) or a valence electron (7). The energy released by these "down" electrons is used to accelerate the "up" electrons out from the host atom (4, 6, and 8) which may be core (4) or valence electrons (6 and 8). These three processes are described by the movements of the "down" and "up" electrons as follows: core hole filled by core electron to produce a core Auger electron (ccc), core hole filled by core electron to produce valence Auger electron (ccv), or core hole filled by valence electron to produce valence Auger electron (cvv).

The discussion of escape depth for XPS applies to that of AES. As such, AES is a surface sensitive tool showing a probe depth of 50 to 100 angstroms depending upon the material.

Energy Balances - XPS

The parameter related to the chemical environment of the host atom is the binding energy which is calculated from the kinetic energy of the photoelectron. If the photoelectron was ejected from a gas molecule, the changes in the vibrational and rotational energy are small during the photoemission; thus, the binding energy is just the difference between the energy of the incident radiation and the kinetic energy of the photoelectron:

$$E_b = h\nu - E_k \qquad\qquad (5 - 1)$$

The binding energy is usually referenced to an electron at rest in a vacuum. However, for solid samples residing a sample holder in electrical contact with the spectrometer this reference energy is not appropriate. The sample and spectrometer will attain equilibrium between the respective Fermi levels (see Fig. 5 - 8). The energy of the ejected electron must be independent of the manner in which it is measured. The kinetic energy measured by the spectrometer, E_{kin}, when added to the work function of the spectrometer, W_{sp}, must equal the total energy of the photoelectron, E_t. E_k is the observation and W_{sp} is usually constant from one experiment to another. Thus, the Fermi level, E_f, and the total energy of the photoelectron are related by

$$E_t = E_f + W_{sp} + E_k \qquad (5 \text{ - } 2)$$

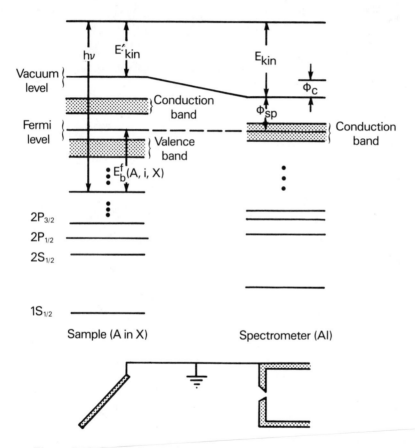

Figure 5 - 8 Energy Level Diagram for a Photoelectron Spectrometer

Reprinted from reference 10, p. 179 by courtesy of Marcell-Dekker, Inc.

For the solid, a similar balance may be written for the binding energy of the photoelectron, E_b. This energy is the difference of the Fermi level, E_f, and the energy level before emission, E:

$$E_b = E_f - E \qquad (5 - 3)$$

The total energy of the electron may be expressed as the sum of the incident radiation and the energy level before emission:

$$E_t = h\nu + E \qquad (5 - 4)$$

Equating the two expressions for the photoelectron total energy, Eqs. (5 - 2) and (5 - 4), gives the following:

$$E + h\nu = E_f + W_{sp} + E_k \qquad (5 - 5)$$

However, the Fermi level minus the electron energy before emission, $E_f - E$, is just the binding energy, E_b. Therefore, Eq. (5 - 5) becomes

$$E_b = h\nu - W_{sp} - E_k \qquad (5 - 6)$$

This binding energy is referenced to an electron at rest residing in the Fermi level of the solid. It must be referred to the energy of an electron at rest in a vacuum using the definition of the sample work function.

$$W_s = E_v - E_f \qquad (5 - 7)$$

and

$$E_{bv} = E_b + E_v - E_f = E_b + W_s = h\nu + W_s - W_{sp} - E_k \qquad (5 - 8)$$

Energy Balances - AES

The energy balances for the AES process are naturally more complicated than those for XPS because it is a two-electron process rather than a one-electron process. Complications to the energy balance for AES include the dependence of the binding energies upon the presence of core holes in the substance, initially (such as p-type semiconductors) and final state multiplet interactions. Neglecting these complications for the moment, we may write the binding energies for a ccc process involving K, L, and M electrons as

$$E_{kin} + W_{sp} = E_b(K) - E_b(L) - E_b^*(M) \qquad (5 - 9)$$

Again, E_{kin} is the observed kinetic energy of the ejected electron from the M shell of the host atom; $E_b(K)$, $E_b(L)$, and $E_b^*(M)$ are the binding energies of the electrons residing, initially, in the K, L, and M shells. The asterisk on the E_b for the M shell means that this energy is for an M electron residing in the field where the K electron has been ejected to form a hole deep in the core. The reader is referred to Reference 5a for a comprehensive treatment of the AES energy balances.

Apparatus

The apparatus for AES and XPS are similar; thus both will be described together. An ultrahigh vacuum is necessary for two reasons. The excited electrons can be captured by gaseous atoms/molecules as well as those in the sample. Thus, the mean free path of the gas molecules should be greater than the dimensions of the spectrometer to ensure that the excited electrons encounter the fewest number of background gas molecules. Secondly, the ultrahigh vacuum is needed to minimize the rate of surface contamination by residual gas in the spectrometer. Usually the sample is cleaned of surface contaminant molecules by heating and argon sputtering. Immediately, upon completion of the sputtering, residual gas molecules strike the surface to become chemisorbed. As the layer builds, the XPS/AES spectrum will take on the attributes of the chemisorbed layer. Once the contaminant layer exceeds the mean escape depth, the electron spectrum is characteristic of the residual gas rather than the sample. At ultrahigh vacuum conditions, the effective analysis time is on the order of hours rather than seconds if the vacuum produced was only a microatmosphere.

A sample holder is required to position the sample(s) properly. Usually multiple samples are positioned by a carousel (Fig. 5 - 9) and each sample is rotated into registry with the incident radiation. Sources for each radiation must be mounted so as not to interfere with the kinetic energy analyzer. Several designs of the energy analyzer have been used, each employing the same principle. The energy of the electron is related to the trajectory taken upon entering an electric field. The path depends upon the kinetic energy and the strength of the field. The most popular analyzer is the cylindrical mirror analyzer (CMA). Here the ejected electrons are made to pass through slits cut in a plate at the end of a cylindrical tube as the electric field surrounding the tube is changed. The resolution of the CMA is increased for a multiple pass trajectory of the electrons. Those electrons successfully negotiating the tortuous path encounter an electron detector.

Applications - XPS

Several examples are prudent to document the success of the XPS also known as ESCA (Electron Spectroscopy for Chemical Analysis). Figure 5 - 10 shows the XPS spectrum of the sulfur atoms in the thiosulfate ion.[6] Two types of sulfur are present in this species: ligand sulfur which is uncharged and the positively charged central ion. These data show the relaxations are resolved by 6 eV. More subtle changes in atomic environment for uncharged species are easily revealed by the XPS

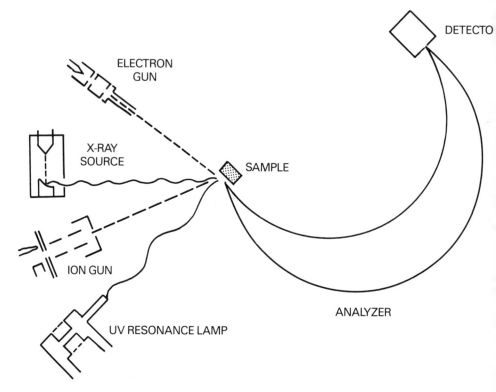

Figure 5 - 9   Schematic Diagram for XPS/AES Apparatus

Left panel: Used with permission from *Experimental Methods in Catalytic Research*, Vol. III., Ed. by R. B. Anderson and P. T. Dawson (New York: Academic Press, Inc., 1976), p. 50.
Right panel: Used with permission from *Methods of Surface Analysis*, Vol. 1., Ed. by A. W. Czanderna (The Netherlands: Elsevier Scientific Publishers, B. V., 1975), p. 170.

technique. Consider Figure 5 - 11, the XPS spectrum of the carbons in ethyl trifluoroacetate, where four distinct peaks are shown for the four different types of carbon in the molecule.[6, 7] Above the spectrum is the representation of the molecule with each type of carbon directly above the XPS peak. The carbon bound to the three fluorine atoms shows a peak at 1188 eV of kinetic energy, and so on. The XPS is not confined to analysis of solid samples as it can also be used to describe the sorption of gases such as carbon monoxide onto different energy sites, Figure 5 - 12. The energy of CO bound to high and low energy sites of tungsten is readily apparent; whereas the clean W surface shows no relaxations in this part of the electron spectrum.[8]

XPS is a valuable tool for determining the surface oxidation state of a

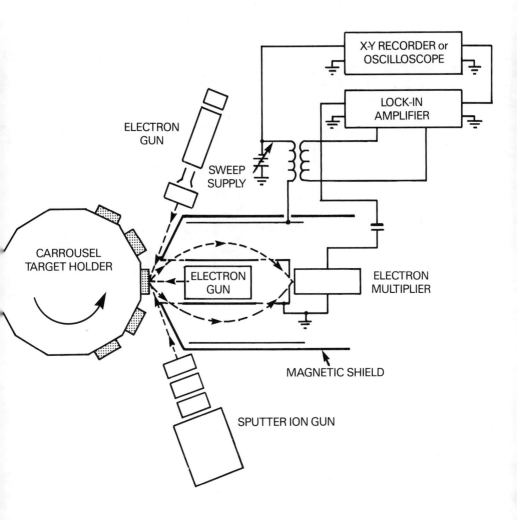

catalyst. In recent studies, XPS has shown the "working state" of some commercial catalysts to be quite different from the bulk oxidation state. Consider the XPS spectra of oxygen and iron (Fig. 5 - 13) in an iron foil sample.[9] At room temperature, the surface of the foil is oxidized as shown by the strong O 1s peak and inferred by the multiple iron peaks. However, these data show how the surface resists reduction in hydrogen at conditions which would easily reduce the bulk sample. Even at 830°C a small oxygen peak is apparent. These results have led scientists to reformulate the theories regarding the stability of the surface phase and to recognize the active phase of the catalysis may not be that of the bulk compound. Figure 5 - 14 documents further the XPS spectra of nickel single crystal (a), nickel foil for which the surface has been sanded (b), nickel foil unsanded (c), and NiO.[10]

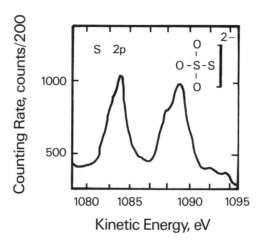

Figure 5 - 10 XPS Chemical Shifts of Thiosulfate Ion

Used with permission from K. Sieghbahn et al., *ESCA: Atomic, Molecular, Solid State Structure Studied by Means of Electron Spectroscopy* (Uppsala: Almqvist & Wiksells, 1967), p. 23.

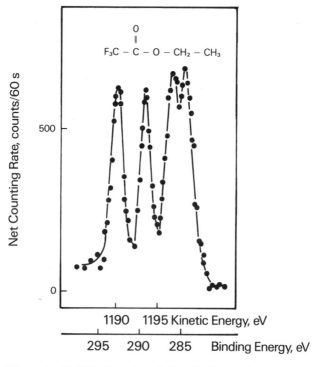

Figure 5 - 11 XPS Chemical Shifts of Ethyl Trifluoroacetate

Used with permission from K. Sieghbahn et al., *ESCA: Atomic, Molecular, Solid State Structure Studied by Means of Electron Spectroscopy* (Uppsala: Almqvist & Wiksells, 1967), p. 21.

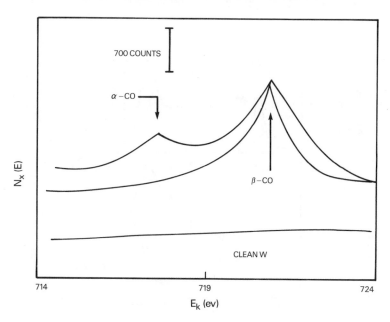

Figure 5 - 12 XPS Spectra of Chemisorbed CO on Tungsten

Oxygen and iron XPS spectra as a function of reduction temperature. Used with permission from T. E. Madley, J. T. Yates, and N. E. Erickson, *Chem. Phys. Lett.*, **19**, (1973), p. 487.

Applications - AES

As with XPS, AES has enjoyed great success in describing the surface composition of solids, such as heterogeneous catalysts. Unlike XPS, AES develops detailed, two-dimensional images of the surface to show element incorporation at some specific surfaces. These images are formed by "rastoring" the incident electron beam to activate the Auger process. Since the time scale of the rastoring process is much longer than that of the atomic Auger processes, an element map of the surface may be developed and presented on a CRT screen.

AES has been used in conjunction with LEED and work function measurements to describe the adsorption of noble gases to single crystals of metals. Consider Figure 5 - 15 showing the Xe Auger intensity (peak-to-peak), the Xe overlayer LEED intensity, and the change in the work function for Xe adsorption on Pd {100}.[11] These data show the Xe Auger peak-to-peak amplitude to increase with exposure until six Langmuirs (L) is attained. At the same exposure, the work function attains a minimum and remains constant for further exposures greater than 6 L. At the point where the Xe Auger peak amplitude levels off, the LEED overlayer intensity changes abruptly indicating that saturation of that surface layer has been attained. The linear increase of the Xe Auger intensity with exposure implies a constant sticking probability up to saturation.

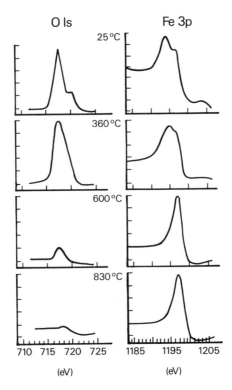

O ls   Fe 3p

25 °C

360 °C

600 °C

830 °C

710   715   720   725   1185   1195   1205

(eV)   (eV)

Figure 5 - 13 XPS Spectra of Iron Foil upon Reduction in Hydrogen

Used with permission from C. S. Fadley and D. A. Shirley, *Phys. Rev. Lett.*, **14**, (1968), p. 980.

X-ray Absorption in Solids (EXAFS)

Recent advances in physics have allowed the practical application of a new technique to the study of the chemistry of very small particles (10 angstroms and smaller). This technique, extended X-ray absorption fine structure (EXAFS), describes the coordination number and radii of the first few coordination spheres for very small particles.[12a, 12b, 12c] Unlike X-ray diffraction, acceptable EXAFS spectra are possible for particles showing order of no more than two coordination spheres of atoms/ions. The obvious applications include highly dispersed noble metal catalysts and ion-exchanged zeolite catalysts. This section will detail the atomic processes, mention the necessary equipment, and show a few examples of the applications.

The absorption of X-radiation by atoms is well documented to occur at very sharp energy levels depending upon the material and the wavelength of the incident X-rays. Figure 5 - 16 shows such an absorption "edge" for a copper crystal. Notice

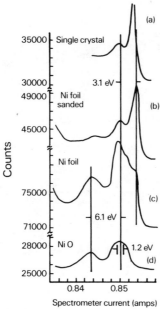

Ni (2p$_{3/2}$) photoelectron lines from NiO and nickel metal.

Figure 5 - 14 XPS Spectra of Ni Foil upon Reduction and Reoxidation

Reprinted from Reference 10, p. 179 by courtesy of Marcell-Dekker, Inc.

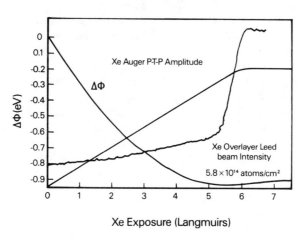

Xe Exposure (Langmuirs)

Figure 5 - 15 AES Data for Xe on Pd (100)

Plots of Auger amplitude, LEED beam intensity, and work function as a function of Xe exposure. Used with permission from P. W. Palmberg, *Surface Science,* **25**, (1971), p. 598.

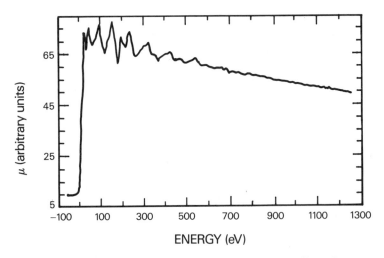

ENERGY (eV)

Figure 5 - 16 X-ray Absorption Edge for Copper Crystal

K-shell mass absorption coefficient of copper versus photon energy of X-ray. Used with permission from E. A. Stern, *Phys. Rev. B.* **10**, No. 8, (1974), p. 3027.

the fine structure in the absorption coefficient beginning at the edge and continuing for up to 400 eV above the edge energy, taken to be zero at the edge. This fine structure is explained to be the result of constructive/destructive interference of the excited electrons (shown here as wave functions) radiating from the neighboring atoms/ions in the crystal (see Fig. 5 - 17). These interference patterns influence the absorptivity of the crystal for the X-radiation due to fluctuations in the wave functions; hence the probability density of the electrons. Recall for any incident radiation the attenuation of the beam is just the integral of the electron wave functions (including the complex conjugate) over the electric field vector for the host atom. Back-scattering of the electron wave functions clearly influences the outcoming wave functions; hence the absorptivity of the host atom. Thus, the fine structure in the absorption edge is just the interferogram of the back-scattered waves, Figures 5 - 18 and 5 - 19. While these interference spectra may yield useful information, it is convenient to transform the data from interference space (I-space) to absorbance space (A-space) using the techniques of Fourier transforms. Recall the absorption coefficient in A-space is related to the data in I-space by the transform equation shown here:

$$A(r) = \int_{-\infty}^{\infty} (I(k)\ exp[-ikr])dk \qquad (5 - 10)$$

Thus, the I-space data may be convolved into A-space yielding the more familiar absorption spectra. It is clear from a consideration of Figures 5 - 18 and 5 - 19 that the quality of the absorbance data depends upon the Fourier transform. The data of

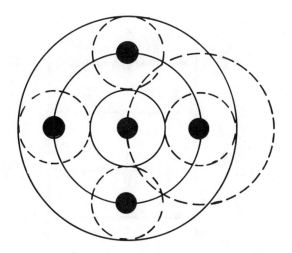

Figure 5 - 17 Model for Scattering of an S-Wave in a Metal Crystal

Hatched circles represent position of atoms. The excited electric state is centered about the center atom. Solid circles represent crest of outgoing part of the electron state. Surrounding atoms diffract the outgoing part as shown by dotted circles. Used with permission from E. A. Stern, *Phys. Rev. B.* **10**, No. 8, (1974), p. 3027.

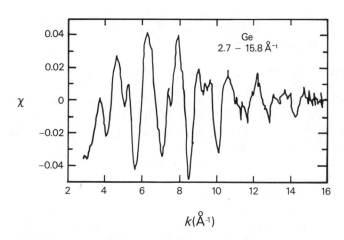

Figure 5 - 18 EXAFS Interferogram of Germanium Crystal

x(k) is X-ray absorption coefficient, k is wavelength of radiation in reciprocal angstroms. Used with permission from E. A. Stern, D. E. Sayers, and F. W. Lytle, *Phys. Rev. B.* **11**, No. 12, (1975), p. 4836.

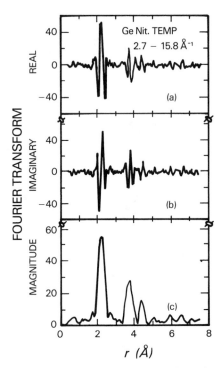

Figure 5 - 19 EXAFS Interferogram and FT Data for Ge at 77 K

Real part (a), imaginary part (b), and magnitude (c) of Fourier Transform of data in Figure 5 - 18. Used with permission from E. A. Stern, D. E. Sayers, and F. W. Lytle, *Phys. Rev. B.* **11**, No. 12, (1975), p. 4836.

the later publications show higher quality transformed data in that the peaks are very obvious. The position of the peaks give the radii of the first, second, and so on coordination spheres; whereas the area of the curves gives the number of atoms in each sphere. For the first sphere, this number is just the coordination number of the host atom/ion. The strength of this technique allows for the coordination numbers (CN) of several ions to be evaluated at once just by scanning the appropriate edge for each ion. The EXAFS technique shows higher quality data for those atoms/ions having larger atomic numbers (Z). Thus, atoms such as carbon and oxygen give poorer EXAFS data than the transition metal atoms and the noble metals.

Subsequent refinement to the data reduction techniques allows for easy assignment of peaks to the radii of the cordination spheres (see Figs. 5 - 18 and 5 - 19). Here for germanium, the interference pattern (Fig. 5 - 18) is reported as the normalized absorption coefficient versus the energy of the radiation in reciprocal angstroms. Fourier transform of these data gives the probability distribution in Figure 5 - 18c. The maxima at radii equal to 2.0, 3.9, and 4.2 angstroms as the indication of the radii of the first three coordination spheres. The Fourier transform of the EXAFS data for crystalline copper is shown in Figure 5 - 19c; note the maxima at 2.3, 3.5, and 4.1 angstroms. The change in the conditions of the analysis (i.e., lower temperature) causes a two-fold increase in the magnitude of the absorbance.

Apparatus

The EXAFS technique may be applied to amorphous as well as crystalline metals and some insulating oxides. Consider Figure 5 - 20 for the EXAFS data for crystalline and amorphous samples of germanium oxide. The spectra (c and d) for radii less than 2 angstroms are very similar, corresponding to the first coordination sphere of oxygen atoms; whereas the second coordination sphere of Ge atoms is less obvious in the amorphous sample compared to the crystalline sample.

The requirements of an ultrahigh vacuum are not neccessary for all EXAFS equipment. In fact, the first EXAFS studies were performed on an automated Bremstrahlung device for which the energy of the X-ray beam was increased stepwise as the absorption of the beam was measured. This method proved to be clumsy and the quality of the EXAFS data was poor. Intense, "white" radiation was necessary to facilitate fast acquisition of high quality data. Synchrotron radiation from storage rings provided the necessary source of X-rays. However, access to the storage ring is limited and costly. These rings are normally located in national laboratories (e.g., Brookhaven National Synchrotron Light Source at the Brookhaven National Laboratories).

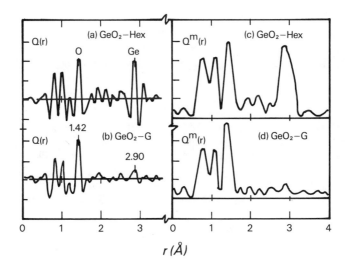

Figure 5 - 20 FT EXAFS Data of Crystalline & Amorphous Germanium Dioxide

Real part (a and b) and magnitude (c and d) of FT of crystalline (Hex) and amorphous (G) GeO$_2$. Structure at 1.42 angstroms is due to first neighboring oxygen atoms. Structure at 2.9 angstroms to second neighboring atoms (Ge). Used with permission from E. A. Stern, D. E. Sayers, and F. W. Lytle, *Phys. Rev. B.* **11**, No. 12, (1975), p. 4836.

Investigators acquire time on these light sources and with the cooperation of personnel at the laboratory, investigator equipment is interfaced with that of the laboratory. In this case, ultrahigh vacuum equipment is necessary to make the attachment to the storage ring. This apparatus provides for positioning of the sample in the beam as well as the necessary windows to allow the radiation access. For the case of catalyst samples, antechambers for conditioning of the samples are necessary. Detection equipment of the type to sense X-radiation is necessary along with sophisticated computer hardware and software to perform the Fourier transform of the raw data.

Conclusions

This brief overview of a few surface science tools illustrates the power of each technique to describe the surface of a solid and the respective limitations of each. No one tool is a panacea for the catalyst scientist. The challenge is to use only those tools which will provide the most cost-effective data. In most instances the design of the experiment may require as much time as the data gathering.

REFERENCES

1a. Kemball, C., *Bull. Soc. Chim. Pays-Bas Belg.*, **67**, (1958), p. 373.

1b. Bond, G. C, *Catalysis by Metals* (London: Academic Press, Inc., 1960).

2a. May, J. W., *Adv. Catal.*, **21**, (1970), p. 151.

2b. Armstrong, R. A., *Experimental Methods in Catalytic Research,* Vol. III, Ed. by R. B. Anderson and P. T. Dawson (New York: Academic Press, Inc., 1976).

2c. Pendry, J. B., *Low Energy Electron Diffraction, The Theory and Its Applications to Determination of Surface Structures* (New York: Academic Press, Inc. 1974).

3. Maire, G., P. Bernhardt, P. Legare, and G. Lindauer, *Proc. 7th Int. Vac. Cong. and 3rd Int. Conf. Sol. Surfaces*, Vienna, (1977), p. 87.

4. Germer, l. and A. U. MacRae, *J. Appl. Phys.*, **33**, (1962), p. 1382.

5a. Hagstrum, H. D., J. E. Rowe, and J. C. Tracy, *Experimental Methods in Catalytic Research*, Vol. III, Ed. by R. B. Anderson and P. T. Dawson (New York: Academic Press, Inc., 1976).

5b. Fadley, C. S., S. B. M. Hagstrom, M. P. Klein, and D. A. Shirley, *J. Chem. Phys.*, **48**, (1968), p. 3779.

6. Siegbahn, K. et al., *ESCA: Atomic, Molecular, and Solid State Structure Studied by Means of Electron Spectroscopy* (Uppsala: Almqvist and Wiksells, 1967).

7. Sieghbahn, K. et al., *ESCA Applied to Free Molecules* (Amsterdam: North-Holland, Publ., 1969).

8. Madey, T. E., J. T. Yates, and N. E. Erickson, *Chem. Phys. Lett.,* **19**, (1973), p. 487.

9. Fadley, C. S. and D. A. Shirley, *Phys. Rec. Letters,* **14**, (1968), p. 980.

10. Delgass, W. N., T. R. Hughes, and C. S. Fadley, *Catal. Reviews,* **4**:2, 179, Marcell Dekker, Inc., (1970).

11. Palmberg, P. W., *Surface Science,* **25**, (1971), p. 598.

12a. Lee, P. A., P. H. Citrin, and B. M. Kincaid, *Rev. Mod. Phys.,* **53**, (1981), p. 769.

12b. Lytle, F. W., G. H. Via, and J. H. Sinfelt, *Synchrotron Radiation Research,* Ed. by H. Winick and S. Doniach (New York: Plenum Press, Inc., 1980), p. 401.

12c. Sinfelt, J. H., G. A. Via, and F. W. Lytle, *Catal. Rev.,* **26**, (1984), p. 81.

CHAPTER 6

EXPERIMENTAL REACTORS

Reaction mechanisms can only be verified by experiment. Often, the challenging part of a research program is the design of the experiment to allow the clearest description of the phenomenon of interest. Consider now the study of an intrinsic reaction mechanism for a surface catalyzed reaction. The design of a kinetics experiment must allow for the measurement of reaction rates, partial pressures, and reaction temperatures. Moreover, these variables must be changed in such a manner so as to extract reaction orders, rate constants, and Arrhenius parameters. Proper reactor design must be considered to eliminate the disguise effects of external/internal mass transfer mechanisms. Finally, reaction conditions must be established to minimize the homogeneous reaction rate.

Choice of Reactor Type - Integral versus Differential

The first choice of experiment design is that of reactor type: an integral reactor or differential reactor. By integral reactor we mean such a reactor where finite conversions occur within a reactor to create significant gradients within the reaction volume. Consider the integral plug flow reactor containing W_g of catalyst bed in volume V with inlet temperature, T_{in}, inlet reactant concentration, C_{in}, and inlet volumetric flowrate, v. The outlet conditions are designated with subscript *out*. Thus, the relation between the reaction rate, r, and the design parameters are

$$V/v = \int_{Cin}^{Cout} dC/(-r) \tag{6-1}$$

Since the purpose of the experiment is to determine the function, r, the integral plug flow reactor is a poor choice, especially if there are temperature gradients within the reaction volume.

The integral backmix reactor under the same conditions shows an algebraic relationship between the reaction rate and the design variables.

$$V/v = (C_{in} - C_{out})/(-r)_{out} \tag{6-2}$$

which may be rearranged to give an explicit expression for the reaction rate:

$$(-r)_{out} = (v/V)(C_{in} - C_{out}) \tag{6 - 3}$$

Thus, the reaction rate, evaluated at the outlet conditions, C_{out} and T_{out}, may be easily measured. It must be shown, however, that the reaction volume is uniform. If the proper analysis shows uniform mixing then the integral backmix reactor *may* warrant further consideration as the potential experimental reactor. It must be shown that external mass transport mechanisms do not limit the observed reaction rate.

Tests have been developed to diagnose the combined ills of incomplete mixing and external mass transport effects. Consider the effects of incomplete mixing upon the observed reaction rates. Levenspiel[1] shows one *model* of incomplete mixing in a CSTR as the recycle plug flow reactor described by the following equation:

$$V/v = -(R + 1) \int_{Ca(in)}^{Ca(f)} dC/(-r) \tag{6 - 4}$$

where

$\quad C_a(f) \quad$ = concentration of reactant leaving the fixed bed
$\quad C_a(in) \quad$ = concentration of reactant entering the fixed bed
$\quad\quad\quad$ = $\{R(C_a(f)) + C_a^* \}/\{R + 1\}$
$\quad C_a^* \quad$ = inlet concentration of reactant to the system

It can be shown as R approaches infinity, then Eq. (6 - 4) approaches Eq. (6 - 2). Thus, for increasing values of internal mixing of the PFR, the observed rates will change in such a way as to mimic that of a CSTR rather than a PFR. This equation is often used to model the recycle reactor.

The external mass transport relation is

$$r = k_m(C_{bulk} - C_{surface}) \tag{6 - 5}$$

where

$$k_m = f(Re, Sc) \tag{6 - 6}$$

In the typical backmix reactor, the Reynolds number (Re), hence k_m, may be increased by increasing the mixing rates. Thus, the observed reaction rates in the transport limited regime will change with mixing in the reactor. The diagnostic for CSTR is to vary the impeller speed or internal recycle fluxes while holding the overall reaction conditions steady. The proper mixing in the vessel will established when further increases in the impeller speed produce no further changes in the observed reaction rates. The same type of diagnostic can be developed for the PFR. Here the Reynolds number is changed at constant space-time by increasing the bed volume of catalyst, V, in proportion to the increase in volumetric flowrate, v, so as to keep the ratio, V/v, constant as the volumetric flowrate (hence the Reynolds number) is increased. As with the CSTR, the proper volumetric flowrate and catalyst volume will be determined when further increases in the volumetric flowrate at constant V/v produce no further changes in the observed reaction rate. This is the classical test for external transport limitations.

Integral, Gradientless Reactors

The batch recirculation reactor and the gradientless flow reactors contact a bed of heterogeneous catalyst particles with a reacting fluid and allow for the gradients within the reaction volume to be minimized in such a way that the measured reaction rates are not disguised by heat and mass transport effects. These transport effects are usually denoted as interparticle effects to discriminate from the effects of internal pore diffusion and reaction which are called intraparticle effects. Equations (6 - 5) and (6 - 6) show the interparticle effects as functions of the local hydrodynamics and give rise to differences between the bulk and surface concentrations and temperatures. Since the surface conditions cannot be measured directly, we must relate them to the bulk phase properties which can be measured. For the gradientless reactors, the diagnostic for elimination of interparticle effects is to measure the reaction rates at fixed values of the severity variables (i.e., space-time in a flow reactor) for increasing values of the variables characterizing the hydrodynamics (e.g., the Reynolds number, Re). This same diagnostic may be applied to integral reactors to characterize interparticle effects only; axial and radial gradients within the integral reactor volume cannot be diagnosed with this technique.

Batch Recirculation Reactors

Kinetics and mechanism studies of heterogeneously catalyzed reactions are often conducted in Pyrex, batch recirculation reactors. The components of these systems (Fig. 6 - 1) are the gas chromatograph sample loop, the catalyst holder/furnace assembly, the mixing volume, and the pump. Usually, these reactors are attached to a glass vacuum system serviced by mechanical/diffusion pumps. These reactors show several characteristics desirable for kinetics/mechanism studies. Isothermal reactor beds may be accomplished easily by reducing the catalytic charge to very small amounts (25 to 50 mg) without fear of bypassing. Excellent mixing of gases is possible without sophisticated design. The small reaction volumes, usually less than 1 liter, allow the use of expensive reagents such as deuterium labeled compounds. Product poisoning studies are accomplished with ease. All these factors contribute to make the recirculation reactor a powerful research tool.

The design of the reactor is simple; however, some insight into the problem is possible by a simple model. We reported[2] such a model as given in Figure 6 - 2. Here, we modeled the system as a two-element, ideal reactor system connected by conduits of negligible volume. All regions of maximum mixing are given by volume V_s; whereas all regions of minimum mixing are lumped together as volume, V_p. The catalyst bed of volume, V_c, weighs W_c grams. The concentration as a function of time is monitored at point M. The gases are recirculated at a constant rate of Q and there is no reaction outside of the catalyst bed. Within the catalyst bed the unsteady concentration-time behavior is negligible and we assume first-order kinetics.

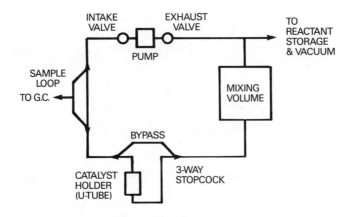

Figure 6 - 1   Schematic of a Laboratory, Batch Recirculation Reactor

Used with permission from M. G. White, O. Bensalem, and W. R. Ernst, *Chem. Eng. J.,* **25,** (1982), p. 223.

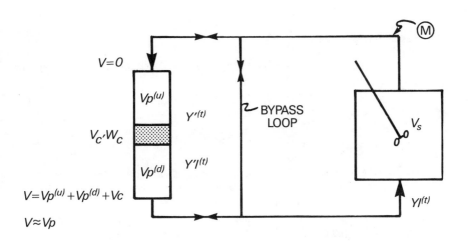

Figure 6 - 2   Two-Element Model of a Batch Recirculation Reactor

Used with permission from M. G. White, O. Bensalem, and W. R. Ernst, *Chem. Eng. J.,* **25,** (1982), p. 223.

The model equations for each element were as follows:

$$V_s: \quad d[y(t)]/dt = [y_1(t) - y(t)]/t_s \tag{6-7}$$

$$y(0) = 1 \tag{6-8}$$

$$V_p: \quad (dy/dt) + Q(dy/dV) = 0 \tag{6-9}$$

$$V_p(u): \quad y(t = 0, V = V_p(u)) = 0 \tag{6-10}$$

$$y(t = t, V = 0) = y(t) \tag{6-11}$$

$$V_p(d): \quad y(t = 0, V = V_p + V_c) = 0 \tag{6-12}$$

$$y(t = t, V = V_p + V_c) = y_1(t) \tag{6-13}$$

Across the catalyst bed, the plug flow design equation for first-order kinetics may be written as

$$d \ln (y) = (-k/Q)dW_c \tag{6-14}$$

These equations were solved to give

$$y(\theta) = \exp[-\theta(1 + a)(1/b)] \sum (1/i!)(\theta(1 + a) - ia)^i \exp[i(a - b)] \tag{6-15}$$

where

$$nab/(1 + a) \leq \theta \leq (n + 1)ab/(1 + a)$$
$$\theta = kW_c t/(V_p + V_s)$$
$$a = t_p/t_s$$
$$b = kW_c/Q$$
$$t_p = V_p/Q$$
$$t_s = V_s/Q$$

The results (Fig. 6 - 3) are conveniently represented as a dimensionless conversion rate, $d[\ln y(\theta)]/d\theta$ for parametric values of a. Figure 6 - 3 gives this dimensionless rate of conversion as a function of the circulation parameter, b^{-1}. For values of $b^{-1} > 25$, the model predicts a reactor performance within 2% of that for the perfectly mixed reactor. In other words, the rate constant determined by this reactor assuming perfect mixing will be in error by no more than 2% if it can be shown that b^{-1} is greater than 25. An unexpected result is the *increased* reaction rate as the fraction of unmixed volume (a) increases. These results are summarized in Figure 6 - 4 where reactor efficiency is defined as the ratio of actual performance to that predicted for the perfectly mixed reactor. At constant value of b^{-1}, increasing a increases the reactor efficiency.

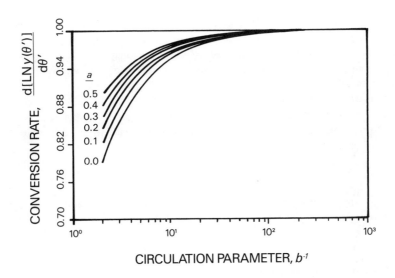

Figure 6 - 3   Effects of Circulation Rate and Unmixed Volume upon Conversion Rate

Used with permission from M. G. White, O. Bensalem, and W. R. Ernst, *Chem. Eng. J.*, **25**, (1982), p. 223.

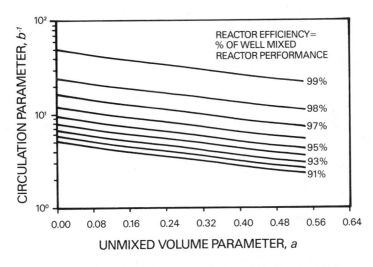

Figure 6 - 4   Correlation of Circulation Rate with Unmixed Volume

Used with permission from M. G. White, O. Bensalem, and W. R. Ernst, *Chem. Eng. J.*, **25**, (1982), p. 223.

This study provides reliable design estimates to build a recirculation reactor. Given an estimate of the reaction rate parameter, kW_c, the desired reaction volume, and the percent unmixed volume, $a/(1 + a)$, then the required circulation rate, Q, may be calculated to ensure the rate constants will meet a prescribed tolerance of that for the perfectly mixed reactor. Higher recirculation rates ensure the measured reaction rates are kinetically controlled.

The properly designed recirculation reactor allows for simple data reduction procedures to be used to generate rate data. For the perfectly mixed batch reactor the change in number of moles of species i

$$(1/\nu_i)[dN_i/dt] = (r_i/\nu_i)V \qquad (6 \text{-} 16)$$

where ν_i is the stoichiometric coefficient and V is the reaction volume. If we define the intensive reaction rate, r, as (r_i/ν_i) then

$$r = (1/V)(1/\nu_i)(dN_i/dt) \qquad (6 \text{-} 17)$$

where N_i is the molar amount of i in the reactor at any time. Thus, kinetic data may be extracted from concentration versus time data in the batch reactor. Particularly useful information is the *initial slope* of these curves. That is,

$$r_o = (1/V)(1/\nu_i)(dN_i/dt)_o \qquad (6 \text{-} 18)$$

for which we may relate to the rate expression

$$r = k(T)g(C_i) \qquad (6 \text{-} 19)$$

$k(T)$ is the familiar rate constant, whereas $g(C_i)$ is the function of concentration. At zero time (i.e., zero conversion) then

$$r_o = k(T_o)g(C_i) \qquad (6 \text{-} 20)$$

where the subscripts denote initial values. The batch recirculation reactor is one type of gradientless reactor normally used in hypobaric applications. For hyperbaric systems, the steel gradientless flow reactors are often used. These reactors are CSTR designs with provisions for containing the catalyst particles in baskets which allow the fluid to contact the catalyst and minimize the external gradients.

Carberry Spinning Basket Reactor[3]

One of the early attempts to fabricate a gradientless, continuous stirred tank reactor (CSTR) for heterogeneously catalyzed reactions was the spinning basket reactor (SBR). The catalyst particles were confined in wire mesh paddles (Fig. 6 - 5) affixed to a spindle which rotates at a desired speed. Two such paddles of two halves each, together with internal baffles, promote the mixing and reduce the external transport gradients between the catalyst and the bulk phase. The Carberry reactor

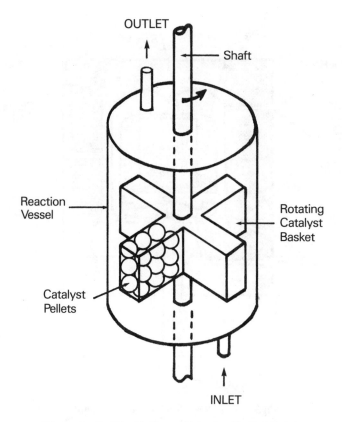

OUTLET

Shaft

Reaction
Vessel

Rotating
Catalyst
Basket

Catalyst
Pellets

INLET

Figure 6 - 5   The Carberry Spinning Basket Reactor

suffers from a poorly defined set of hydrodynamics which do not change in a predictable manner with an increasing rotation speed of the paddles. Finely divided catalysts are difficult to contain in the rotating paddles.

Berty Gradientless Reactor[4]

The Berty reactor is a steel version of the loop reactor which promotes the recycle stream internally in the annular space between the draft tube containing the catalyst particles as a fixed bed and the reactor shell (Fig. 6 - 6). An impeller at the base of the draft tube draws reacting gases through the catalyst bed in piston flow. For increasing impeller speeds, the volumetric flow through the bed increases. This design allows for a definite relationship between the hydrodynamics and the impeller speed. Baffling reduces the velocity component of the gases around the axis of the impeller. Magnetic couplings between the impeller and the electric motor eliminates the need for rotating shaft seals in the reactor. The Berty reactor has enjoyed success in studies of petroleum reactions.[5]

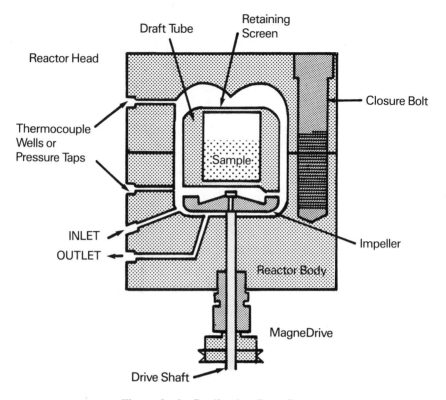

Reactor Head

Draft Tube

Retaining Screen

Closure Bolt

Thermocouple Wells or Pressure Taps

Sample

INLET

OUTLET

Reactor Body

Impeller

MagneDrive

Drive Shaft

Figure 6 - 6   Gradientless Berty Reactor

Modeling of the Berty Reactor has been accomplished using the recycle reactor design equation given by Levenspiel.[1] The fixed catalyst bed contained in the draft tube of the Berty Reactor is pictured as being fed by a recycle stream of volumetric flowrate, R, times the outlet flowrate, v_o, from the reactor system. This recycle stream is the internal recycle generated by the action of the spinning impeller on the gases inside the vessel; the volumetric flowrate of this internal recycle is related to the gas density and the impeller speed. The conversion of reactant across the fixed bed of catalyst is determined by the recycle flowrate, (Rv_o), added to the fresh feed rate, (v_o), and the volume of the catalyst bed, (V). The plug flow design equation applied to this system would appear as follows:

$$(V/(R + 1)v_o) = \int_{Ca(in)}^{Ca(f)} dC/(-r) \qquad (6 - 21)$$

where

$C_a(f)$ = outlet reactant concentration from the reacting system
$C_a(in)$ = inlet reactant concentration to the fixed bed

Chapter 6   Experimental Reactors

The inlet concentration to the fixed bed may be calculated from a simple material balance of the reactant A between the streams entering the bed: the recycle stream of concentration, $C_a(f)$, and flowrate, Rv_o, and the fresh feed of concentration, C_a^*, and flowrate, v_o, for a constant density system. Thus, the concentration to the bed, $C_a(in)$, is

$$C_a(in) = \{RC_a(f) + C_a^*\}/\{R + 1\} \qquad (6 - 22)$$

The asymptotic solutions of Eq. (6 - 21) are useful and instructive. For the trivial case of $R = 0$, the design equation of the plug flow reactor is recovered; however, when R approaches infinity, it can be shown by appealing to L'Hopital's Rule that the recycle design equation approaches that of a CSTR for all kinetics. Thus, the Berty reactor may be modeled with the simple algebraic equations of the CSTR when the recycle fluxes are sufficiently high. The diagnostics for mixing and external transport limitations discussed for the CSTR may be applied to the Berty reactor.

Integral Stratified Reactors

All reactors, whether they be small, laboratory scale reactors or large commercial units are integral stratified reactors. However, the foregoing discussion illustrates the manner by which the gradients inside the integral reactors can be minimized to the point at which the existing gradients do not influence the reaction rates. These gradientless reactors are ideal for kinetics studies and other laboratory scale investigations. The designs of the gradientless reactors are complicated and thus the cost of the reactors is much higher than for the simple tubular flow reactor (TFR). As a laboratory scale reactor, the tubular flow reactor may be suitable if certain reaction conditions are met. For example, nonisothermal operation of the tubular reactor demands that the energy balance be solved simultaneously with the mass balance such that reasonable results are obtained. For highly exothermic reactions, it may be impossible to operate the reactor in the isothermal mode, and runaway conditions may exist over the range of desired operating conditions. Endothermic reactions may be used in tubular reactors with more success. The problem of unsteady-state operation does not exist and isothermal operation may be achieved without much difficulty.

The major objection to using integral stratified reactors in studying kinetics is the difficulty in processing the raw data so as to extract the intrinsic kinetics. When gradients in temperature and concentration both exist in the tubular reactor, a second-order partial differential must be solved for each balance. If the kinetics are nonlinear with temperature-dependent terms coupled with the mass dependent terms (such as Langmuir-Hinshelwood kinetics), the task becomes hopeless. The gradientless reactors were designed to avoid this type of problem in data reduction. The tubular flow reactor, TFR, can be used as a catalyst screening reactor at a fixed set of reaction conditions. In fact, as a qualitative tool the TFR is ideal. It is simple to operate, cheap, and easily lends itself to computer automation. In the capacity of a catalyst screening reactor, the tubular reactor has no equal.

Differential Reactors

The term *differential reactor* describes a mode of operation rather than a particular physical design, although differential reactors may actually be physically smaller than the integral reactor counterparts. For the differential reactor, the operating parameters of catalyst weight and volumetric flowrate are adjusted at the desired reaction temperature such that the reactant fractional conversion is below 10%. For these conditions, the differential expression of the design equation may be approximated by a finite difference equation. Most often, the differential tubular flow reactor (Fig. 6 - 7) is used to gather kinetics data. The data manipulation procedures are described next.

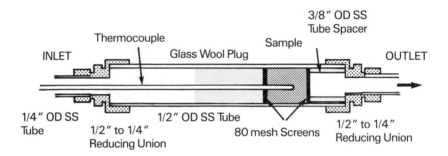

Figure 6 - 7 Microcatalytic, Tubular Flow Reactor

Fixed Bed Microcatalytic Reactor

The differential design equation may be approximated as a finite difference equation as follows:

$$dt_p = -dc/(-r) \rightarrow t_p = (V_p)/v = \Delta C/[(-r)|_{(Cin+Cout)/2}] \qquad (6 - 23)$$

where

t_p = space-time of reactor
V_p = volume of reactor
v = volumetric flowrate through the reactor
ΔC = outlet - inlet concentration of the reactant
$(-r)|_{Ci}$ = kinetic rate expression evaluated at concentration, C_i

In the present case, the rate equation is to be evaluated at the arithmetic average of the inlet and outlet concentration of the reactant. At lower conversions, the average concentration approaches the inlet concentration in the limit of zero conversion.

Thus, Eq. (6 - 23) may be rewritten explicit in the rate as follows:

$$(-r)|_{(Cin+Cout)/2} = C_{in}(f - f_{in})/(t_p) \qquad (6 - 24)$$

where the difference between and outlet concentration has been written in terms of the difference in fractional conversions. If the feed to the reactor is unconverted, then $f - f_{in}$ may be replaced by f. In the limit of zero conversion, this last equation may be rewritten as

$$(-r)|_{Cin} = C_{in}(f)/(t_p) \qquad (6 - 25)$$

Thus, data of finite but low fractional conversion are plotted versus space-time, t_p, and extrapolated to zero conversion to find the initial slope. The corresponding initial reaction rate is found by multiplying this initial fraction conversion rate by the corresponding inlet concentration.

Figure 6 - 8 shows oxygen conversion data versus space-time for several different inlet concentrations of oxygen at one temperature for the oxidative coupling of propylene to hexadiene.[6] Least-squares lines through the low conversion data at each inlet concentration of oxygen define the conversion rate which obviously decreases with increasing oxygen, partial pressure indicating fractional order less than unity for oxygen dependence. The corresponding reaction rate at zero conversion is the product of the conversion rate with the inlet concentration (Fig. 6 - 9).

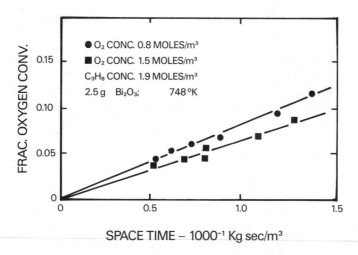

SPACE TIME – 1000⁻¹ Kg sec/m³

Figure 6 - 8   Oxygen Conversion versus Space-Time

Used with permission from M. G. White and J. W. Hightower, *J. Catal.* **82**, (1983), p. 185.

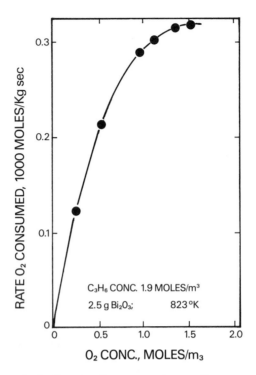

Figure 6 - 9   Reaction Rate versus Oxygen Concentration

Used with permission from M. G. White and J. W. Hightower, *J. Catal.* **82**, (1983), p. 185.

As with integral reactors, the "disguise" mechanisms of transport limitations in the fixed bed reactor are diagnosed by varying the hydrodynamics (here volumetric flowrate) at fixed space-time and observing the fractional conversion. In practice the test is carried out for several different space-times and volumetric flowrates such that comparisons at constant, t_p, but different v can be made (Fig. 6 - 10). Intrapellet transport effects are diagnosed by varying the particle size at constant reactor conditions. Figures 6 - 10 and 6 - 11 show the oxygen conversion and selectivity to desired product as a function of space-time and particle size.[6] The particles from each test were crushed, classified, and returned to the reactor for the subsequent test. The single curve in Figure 6 - 10 clearly shows that the oxygen consumption rate is not affected by the particle size in the range 74 to 3360 microns. However, the selectivity is a function of particle size, as seen by the three separate curves in Figure 6 - 11. Notice that the smallest particles give the highest selectivity.

Versions of the fixed bed, microcatalytic reactor have been used to introduce pulses of the reactant over the catalyst.[7] A flowing stream of inert (He or N_2) is used to establish the mean residence time. These reactors are employed when the bed of

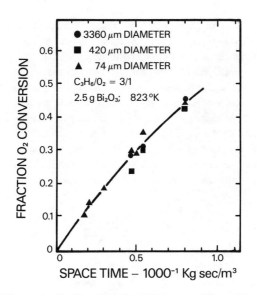

Figure 6 - 10   Particle Size Effects upon Reactivity

Used with permission from M. G. White and J. W. Hightower, *J. Catal.* **82**, (1983), p. 185.

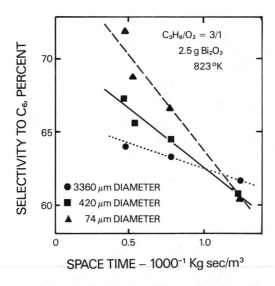

Figure 6 - 11   Particle Size Effects upon Selectivity

Used with permission from M. G. White and J. W. Hightower, *J. Catal.* **82**, (1983), p. 185.

catalyst decays rapidly upon contact with a flowing stream of the reactant or when the reactant is expensive (isotopic tracers). One distinct advantage to the technique is that the reactor effluent may be connected directly to the inlet of a chromatograph. For some applications the catalyst bed may be inserted into the inlet section of the chromatograph; hence the name chromatographic reactor. Highly exothermic reactions have often been studied at nearly isothermal conditions owing to the large heat sink offered by the reactor wall compared to the small bed of catalysts and to the small size of the reactant pulses. One major fault of the pulse reactor is the uncertainty of the space-time caused by broadening of the pulse during the adsorption/desorption over the catalyst. With this uncertainty in space-time is the accompanying uncertainty in any reaction rate data measured with the reactor.

Conclusions

Choosing the correct laboratory reactor is not easy, especially when funds for the research are limited. Weekman[8] reviewed the literature to report on this subject. His article discusses the reactor types presented here and a few not included here. His summary of ratings for different reactor types is presented in Table 6 - 1. As he noted, all the reactors reviewed have at least one poor rating; thus, the researcher is forced to compromise. It is left to the scientist's experience to decide how important the compromise is to the success of the study.

Table 6 - 1   Summary of Reactor Ratings--Single Fluid Phase

Reactor Type	Sampling and Analysis	Isothermality	Residence-Contact Time	Selectivity Disguise-Decay	Construction Problems
Differential	P-F	F-G	F	G	G
Fixed bed	G	P-G	F-G	G	G
Sirred batch	F	G	G	G	G
Stirred-contained solids	G	G	G	G	F-G
Continuous stirred tank	F	G	F	G	F-G
Straight-through transport	F-G	P-F	F-G	G	F-G
Recirculating transport	F-G	G	G	G	P-F
Pulse	G	F-G	P	G	G

G = good, F = fair, P = poor
Reproduced by permission of the American Institute of Chemical Engineers, V. W. Weekman, *Am. Inst. Chem. Eng. J.*, **20**, No. 5, 833 (Sept., 1974).

REFERENCES

1. Levenspiel, O., *Chemical Reaction Engineering* (New York: John Wiley & Sons, 1978).
2. White, Mark G., O. Bensalem, and W. R. Ernst, *Chem. Eng. J.*, **25**, (1982), p. 223.
3. Carberry, J. J., *Ind. Eng. Chem.*, **56**, (1964), p. 39.
4. Berty, J. M., "Reactor for Vapor-Phase Catalytic Studies," paper presented at the 66th Annual Meeting of the A.I.Ch.E., Philadelphia, PA, (1973).
5. Mahoney, J. A., *J. Catal.*, **32**, (1974), p. 247.
6. White, Mark G., Ph. D. thesis, William Marsh Rice University, (1977).
7. Galeski, J. B. and J. W. Hightower, *Can. J. Chem. Eng.*, **48**, (1970), p. 151.
8. Weekman, V. W., Jr., *Am. Inst. Chem. Eng. J.*, **20**, (1974), p. 833.
9. Beckler, R. K., Ph. D. thesis, Georgia Institute of Technology, (1987).
10. Stull, J. O. and M. G. White, *O. E. D.*, Vol 11, (1986), p. 325.
11. Poehlein, S. R. and M. G. White, *O. E. D.*, Vol 12, (1987), p. 55.
12. Oakes, J. D., M. S. thesis, Georgia Institute of Technology, (1983).

PROBLEMS

1. Examine the cyclopropane (CP) data from a batch, recirculation reactor reported by Beckler[9] shown in Figure 6 - 12. The fractional conversion of CP, f, was recorded versus time in three separate experiments for which only the gas

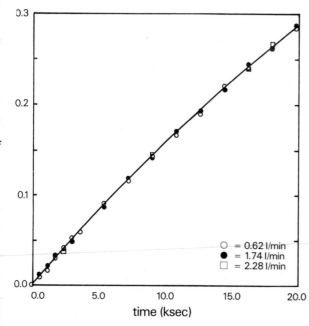

Figure 6 - 12 Cyclopropane Fractional Conversion versus Time for Different Flowrates

CP fractional conversion is plotted against time for three different recirculation gas flowrates at constant temperature (150 C), particle size (100/200 mesh), and initial CP pressure (50 Torr).

○ = 0.62 l/min
● = 1.74 l/min
□ = 2.28 l/min

time (ksec)

recirculation rate (measured in actual liters/min) was changed as follows: 0.62, 1.74, and 2.28 liters/min. For all runs, the temperature was 150°C, the initial CP partial pressure was 50 Torr, the weight of the catalyst was 0.5 g, and the size of these catalyst particles was 100/200 mesh. Does this system experience disguise of the kinetics by external mass transport? ·

2. Consider the CP data at 150°C for varying particle size (40/60, 60/100, and 100/200 mesh) in Fig. 6 - 13 reported by Beckler.[9] For these runs, the CP partial pressure was 50 Torr, the weight of catalyst was 0.5 g, and the gas recirculation rate was 2.28 liters/min. Does this system suffer from internal mass transport limitations? If so, what particle size should be used to avoid these limitations?

3. We reported the effect of particle size upon reaction rates for the simple hydration of potassium superoxide (KO_2) with water vapor[10, 11] This highly exothermic reaction shows the following stoichiometry:

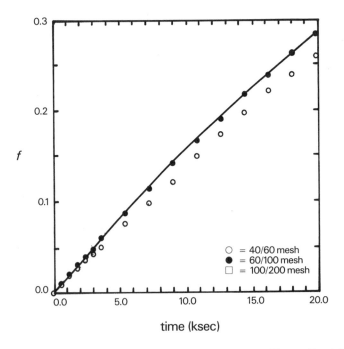

Figure 6 - 13   Cyclopropane Conversion versus Time for Different Particle Sizes

CP fractional conversion is plotted against time for three different particle sizes at constant temperature (150 C), recirculation gas flowrate (2.28 l/min), and initial CP pressure (50 Torr).

Chapter 6   Experimental Reactors

$$2\ KO_2 + H_2O \rightarrow 2\ KOH + 1.5\ O_2 \qquad\qquad \textbf{(6 - 26)}$$

The data were collected at room temperature for pressures between 4.2 and 12 atm with particles between 335 and 2190 microns. Analyze these data (see Fig. 6 - 14) and discuss the particle size effects within the framework of the effectiveness factor theories for isothermal and nonisothermal reactions. See Reference 1 for a discussion of the effectiveness factor theory.

    4. Oakes[12] studied the mixing in a batch recirculation reactor for a step input of tracer. A small dose of the tracer (propylene, 7 ml, 672 Torr at room temperature) was put into the one-liter batch reactor which was filled with He at 540.5 Torr. The He was recirculated at 1.35 actual liters/min. The tracer was admitted and analyzed by gas chromatography at one-minute intervals. These data are shown in Figure 6 - 15 as partial pressure of propylene at time, t, ratioed to the final partial pressure

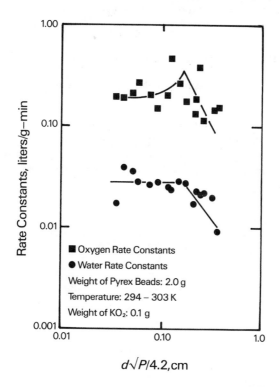

$d\sqrt{P}/4.2$,cm

Figure 6 - 14 Particle Size Effects in KO$_2$ Hydration

Used with permission from S. R. Poehlein and M. G. White, *O. E. D.*, Vol. 12, (1987), p. 55.

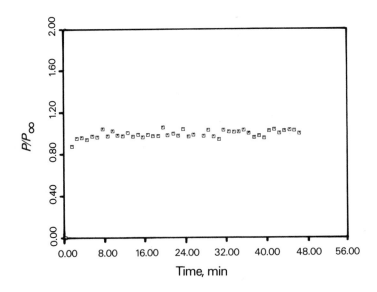

Figure 6 - 15   Step Response for Mixing in a Batch Recirculation Reactor

Used with permission from J. D. Oakes, M. S. thesis, Georgia Institute of Technology (1983).

of propylene in the reactor. Comment on these results as they relate to the degree of mixing in the reactor. Compare these data to the predicted behavior of the model for a recirculation reactor given by Levenspiel.[1] Using Eq. (9 - 56) on page 294 of this reference, extract model parameters for the residence time and number of ideal CSTR's in series.

PART III

REACTION KINETICS

The kinetic theories of homogeneous reactions are reviewed to provide a fundamental background for the heterogeneous reactions. The absolute rate and transition state theories for homogeneous reactions are reviewed to give added insight to the molecular processes occurring during the reactive events. Rate laws for homogeneous and heterogeneous reactions are integrated and linearized. Case studies are developed to illustrate some of the theories.

CHAPTER 7

THE KINETICS OF HOMOGENEOUS REACTIONS

Reaction kinetics specifies the relationship between the observed reaction rates and the reaction temperature, pressure, and species concentrations for homogeneous reactions. Often, for heterogeneous reactions, the type of catalyst will play an important role in determining the reaction rate. In this chapter we will define some important terms, review theories of absolute reaction rates and transition state kinetics, and describe several rate laws.

Definition of Terms

We distinguish between extensive and intensive reaction rates. Extensive reaction rate in units of moles/time is the rate observed for a specific reaction volume or for a specific amount of catalyst. Intensive reaction rates for homogeneous reactions (units $[=]$ moles/volume-time) take into consideration the reaction volume. For heterogeneous reactions, convenient variables to make the rate specific include the weight or volume of catalyst, the surface area of the catalyst, and the active site density. In the last instance the rate per site is called the turnover number, to represent the average rate that an individual reactive site "turns over" to release a product and accept a new reactant molecule.

The reaction mechanism is just a description of those molecular events leading to the desired reaction products and the side products. In the case of heterogeneous reaction, the mechanism includes all adsorption/desorption events plus activation/deactivation events. As an example of a homogeneous mechanism consider the reaction to form hydrogen bromide (HBr) from the diatomic molecules hydrogen and bromine. The first step involves the activation of bromine to form bromine radicals (atoms) which may either react with other bromine molecules in a nonproductive radical transfer reaction or with hydrogen molecules to form the product plus hydrogen radicals. Termination steps involve the recombination of either the like radicals to form the reactants or unlike radicals to form the product. Possible poisons to the reaction are radical scavenging agents such as hydrogen sulfide and butadiene. The mechanism would be described as follows:

Initiation: Br-Br $\Leftrightarrow$ 2Br$\cdot$

Radical Transfer: Br· + Br-Br ⇔ Br-Br + Br·
Br· + H-H ⇔ Br-H + H·
H· + H-H ⇔ H-H + H·
H· + Br-Br ⇔ H-Br + Br·

Termination: 2Br· ⇔ Br-Br
2H· ⇔ H-H
H· + Br· ⇔ H-Br

"Poisoning": Radical· + Scavenger ⇔ ?

The value in having such a mechanism is that reaction rate expressions may be written for each elementary reaction using the theories of absolute reaction rates or transition state kinetics. For example, in the initiation step the rate expression for this reversible reaction is just

$$-r = k([Br_2] - 1/K[Br·]^2)$$

The rate constant for the forward reaction, k, and that for the reverse reaction, k', are ratioed to give the equilibrium constant: $K = k/k'$. The square brackets denote species concentration. If one knows the concentrations of the bromine molecules and atoms and the forward/reverse rate constants, then the net rate of consumption of bromine molecules by the initiation step is determined easily.

Theories of Homogeneous Unimolecular Reactions

Homogeneous reactions are those which occur between molecules/atoms in one phase only. Reactions at phase boundaries are excluded from consideration. In fact, tests for wall-catalyzed reactions are usually performed to show that homogeneous reactions dominate. Let us consider now the classes of reactions as unimolecular, bimolecular, termolecular, and so on. The basis for such classification originally arose from the molecularity of the elementary reactions. That is, unimolecular reactions were thought to involve only one molecule, bimolecular reactions involved two molecules, and so on. However, the latest studies reveal all reactions are at least bimolecular.

Perrin Theory for Unimolecular Decomposition

The early concepts of unimolecular reactions are found in the theory according to Perrin[1] who suggested that the reactions were initiated by the absorption of radiation thereby activating a bond in the reactant molecule. This postulate arises as a natural consequence of the first-order rate laws observed for the unimolecular reactions. This linear rate law,

$$(-1/c) \, dc/dt = k \qquad (7 \cdot 1)$$

shows the fractional reactant conversion rate is clearly independent of total pressure since the isothermal rate constant, k, was observed to be constant over a range of pressures. Thus, all pressure-dependent processes, such as collisions, could not be involved in the rate determining step.

Black body radiation was postulated as the energy source for activation of the reactant bond. The fractional reaction rate was assumed to be a function of the radiation density at the characteristic frequency of the critical bond. The activation energy, E, was just Avogodro's number times the product of Planck's constant, h, and the characteristic frequency, ν. Thus, the population of vibrators having the proper energy, E, at temperature, T, was given by the familiar expression

$$k = k_o \exp(-E/RT) \tag{7 - 2}$$

Unfortunately, attempts to verify this theory were unsuccessful.[2] In one case, N_2O_5 gas at temperatures sufficiently low to suppress the thermal decomposition was irradiated with intense light at the required wavelength of 1.16 microns predicted from activation energy data gathered at higher temperatures. No reaction was observed nor was any light absorbed. The Perrin theory was a failure.

Lindemann Theory for Unimolecular Decomposition Reactions

Subsequent studies, focusing on the low pressure kinetics of the unimolecular reactions, showed that the observed, first-order rate constants changed with total pressure. Thus, the kinetics of the unimolecular reactions appeared to be linear only at "sufficiently" high pressures and showed second-order kinetics at sufficiently low pressures. A collision process must be involved the activation of the reactant. Lindemann[3] proposed collisional activation by a bimolecular process followed by unimolecular decomposition of the activated complex. At high pressures many of the activated complexes may be deactivated by a subsequent collision before the decomposition occurred. Thus, the collision activation is pictured as a quasi-equilibrium process; whereas the decomposition is the irreversible, rate controlling step.

$$A \Leftrightarrow C \rightarrow products \tag{7 - 3}$$

The mechanism is given next for the activation of reactant molecule, A, to the complex A^* via a bimolecular, reversible collision with forward and reverse rate constants, k_1 and k_{-1}.

$$A + A \Leftrightarrow A + A^* \tag{7 - 4}$$

The activated complex then decomposes by a second, irreversible reaction as given next with rate constant, k_2.

$$A^* \rightarrow products \qquad\qquad (7 - 5)$$

We may write the net rate of complex formation via the quasi-equilibrium, bimolecular and unimolecular decomposition reactions as follows:

$$d[A^*]/dt = k_1[A]^2 - k_{-1}[A][A^*] - k_2[A^*] \qquad\qquad (7 - 6)$$

The next step is to invoke the pseudosteady state hypothesis to determine the activated complex concentration in terms of the reactant concentrations and rate constants. This hypothesis states that the complex concentration is very small which is no more than a restatement of what is observed. However, as a consequence of the short lifetimes of the complexes, the net rate of production for the complex $(d[A^*]/dt = 0)$ must be zero at all reactions times greater than the induction period. This hypothesis allows for Eq. (7 - 6) to be solved explicitly for $[A^*]$.

$$[A^*] = (k_1[A]^2)/(k_2 + k_{-1}[A]) \qquad\qquad (7 - 7)$$

Equation (7 - 7) may be combined with the rate law for reaction (7 - 5) and the stoichiometry $-A = A^*$ to give

$$(-1/[A]) \, d[A]/dt = (k_1 k_2[A])/ (k_2 + k_{-1}[A]) \qquad\qquad (7 - 8)$$

If we compare Eqs. (7 - 1) and (7 - 8), the observed rate constant, k, is equal to the right-hand side of Eq. (7 - 8). For high concentrations of reactant, A, the linear concentration term dominates the denominator, thus, the function approaches a pressure-independent constant, $k_1 k_2/k_{-1}$. At low pressures, the zero-order term dominates the denominator; thus, the observed rate "constant" becomes a function of reactant concentration (i.e., pressure dependent).

Physically, Eq. (7 - 8) predicts a process for which the rate limiting steps changes with reactant partial pressure if there are no other species present in high concentrations. As the pressure decreases, the rate of the activation/deactivation steps decreases by the square of the reactant concentration; whereas the unimolecular decomposition rate decreases linearly with the complex concentration. At some characteristic reactant concentration, the rates of activation/deactivation by collisions become equal to that of the decomposition reaction and the collision frequency governs the overall reaction rate.

The pressure dependence of the first-order rate constant at low pressures for the isomerization of methyl isocyanide to acetonitrile for three temperatures is given in Figure 7 - 1.[4] The ratio of the rate constant at pressure, p, to that at high pressure is plotted versus pressure on log-log coordinates. At the high pressure limit each isotherm shows a zero slope indicating no pressure effect on the rate constant; whereas the isotherms approach a slope equal to unity at the low pressure limit as predicted by Eq. (7 - 8).

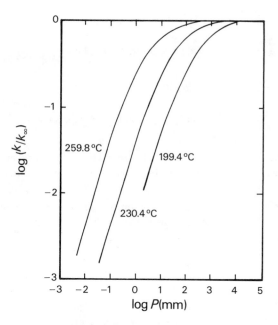

Figure 7 - 1 Pressure Dependence of Unimolecular Rate Constants for the Isomerization of Methyl Isocyanide

For clarity, the 260°C curve is arbitrarily displaced by one log(P) unit to the left in the figure while the 200°C curve is displaced the same distance to the right; actually both of these curves would almost coincide with the 230°C curve. Vertical marks have been placed under the 200°C high pressure points to assist in distinguishing them from the 230°C data. The curves represent the calculated results. Reprinted with permission from F. W. Schneider and B. S. Rabinovitch, *J. Am. Chem. Soc.* **84**, (1962), p. 4215. Copyright 1962, American Chemical Society.

We have not specifically mentioned the effect of inert molecules upon the unimolecular kinetics. If we rewrite the activation/deactivation steps to include such inert molecules, denoted by M, then

$$A + A \Leftrightarrow A + A^* \tag{7 - 9}$$

$$A + M \Leftrightarrow A^* + M \tag{7 - 10}$$

If the concentration of M is much greater than that of A, then Eq. (7 - 10) is the activation process which predominates. Now the net rate of complex formation is given by

$$d[A^*]/dt = k_1[A][M] - k_{-1}[A^*][M] - k_2[A^*] \tag{7 - 11}$$

Using the pseudo steady-state hypothesis we have

$$[A^*] = k_1[A][M]/(k_2 + k_{-1}[M])$$ (7 - 12)

The reaction rate for the decomposition is

$$(-1/[A]) \, d[A]/dt = k_2 k_1[M]/(k_2 + k_{-1}[M]) = k$$ (7 - 13)

Clearly, the observed rate constant, k, is a function of total pressure even when the partial pressure of A is held constant in a mixture of A and M since $[M]$ must change with total pressure.

The effects of inert molecule, M, upon a reaction are represented by the ratio of the rate constants for M present in excess to the rate constant for only A present, k_m/k_a. Using the low pressure limit of Eq. (7 - 13),

$$k_m/k_a = (k_1^M/k_1^A)([M]/[A])$$ (7 - 14)

where k_m, k_a are the observed rate constants. The intrinsic rate constants for activation are k_1^M and k_1^A. Table 7 - 1 shows the results for some of the 102 different inert gases used with the methyl isocyanide isomerization reaction.[5] The atoms and the light, simple molecules show the smallest ratio of the reaction rate constants; whereas the heavy, complicated molecules show the largest ratio of rate constants. These ratios may be corrected for reduced mass (μ) and size effects (σ) as shown in column three of Table 7 - 1. The adjusted ratios show that the least efficient collisions for activation of the isocyanide molecules involve monatomic gases and the most effective collision partners are the polyatomics.

Consider now the equilibrium ratio of the forward to the reverse rate constants for the activation step, K, Eq. (7 - 10). From thermodynamics, this equilibrium ratio is just the product of product concentrations divided by that of the reactant concentration. Since the concentration of inert molecule, $[M]$, appears in numerator

$$K = [A^*][M]/[A][M] = [A^*]/[A]$$ (7 - 15)

and denominator, this equilibrium constant is equal to the fraction of A molecules that have been activated and is independent of M. These results in Table 7 - 1 suggest a small molecule is less effective than a large molecule for activating the complex A^*, and these larger molecules should produce higher concentrations of the complex for longer lifetimes. The longer lifetimes of the complex should afford a greater opportunity for the complex to decompose to the products. Apparently, effective molecules for activation of the complex are those having energy states compatible with those of the activated complex.

Hinshelwood's Modification to the Lindemann Theory

A modification to the Lindemann theory proposed by Hinshelwood[6] abandoned
the hard-sphere collision model for activation in favor of one that allowed
interchange of energy between internal and external modes (e.g., vibrational and
translational). The assumptions were

1. Use classical equations rather than quantum mechanics.
2. The external mode of translation supplies the energy for activation via
   the internal vibration mode as a result of two-body collisions only.
3. The minimum energy for activation, E^*, need not be localized in the
   one bond to be activated. Energy in excess of E^* would not increase
   the probability of reaction.
4. The classical distribution for n vibrators is used for a molecule
   containing N atoms.

The classical distribution function for n vibrators is given by

$$P(E) \, dE = \{(E/kT)^n\}/(n - 1)! \, exp(-E/kT) \, d \ln (E) \qquad (7 - 16)$$

where $\{P(E) \, dE\}$ is the fraction of vibrators having energy between E and $E + dE$.

If we integrate Eq. (7 - 16) from the minimum energy, E^* to infinity, the resultant is the fraction of oscillators having the minimum energy, E^* and greater.

$$f(E^*) = \{(E^*/kT)^{(n-1)}\}/(n - 1)! \, exp(-E^*/kT) \qquad (7 - 17)$$

Recall the earlier discussion of the equilibrium constant, K, for the activation of A to A^*, Eq. (7 - 15); Eq. (7 - 17) gives the fraction of A molecules that have been activated by collision at equilibrium. Thus

$$f(E^*) = [A^*]/[A] = K = k_1/k_{-1}$$

Solving for the forward rate constant, k_1, gives

$$k_1 = k_{-1}f(E^*) \qquad (7 - 18)$$

and for the reverse rate constant, k_{-1}, use the collision frequency for a nonactivated process.

$$k_{-1} = z/[A]^2 \qquad (7 - 19)$$

The collision frequency from the kinetic theory of gases for hard spheres is

$$z/[A]^2 = 2\pi d^2 (kT/\pi m)^{0.5} \qquad (7 - 20)$$

Thus, the forward rate constant, k_1, for the activation step is

$$k_1 = 2\pi d^2 (kT/\pi m)^{0.5} (E^*/kT)^{(n-1)} \, exp(-E^*/kT)/(n - 1)! \qquad (7 - 21)$$

As before, a balance on the net rate of complex formation gives the concentration of the activated complex which is multiplied by the decomposition rate constant, k_2 to yield the rate of reaction (e.g., see Eq. 7 - 8). The decomposition rate constant is assumed to be independent of temperature since the complex has been activated by the collision and no further activation should be necessary for the reaction to occur. The unimolecular rate constant, k, is

$$k = k_2 f(E^*)[A]/(k_2/k_{-1} + [A]) \qquad (7 - 22)$$

Equation (7 - 22) may be linearized by inverting it to yield the following:

$$1/k = 1/\{k_{-1}f(E^*)[A]\} + 1/k_2 f(E^*) \qquad (7 - 23)$$

The data of the first-order rate constant, k, versus concentration of A may be plotted according to the linearizing Eq. (7 - 23) to give the following:

$$intercept = 1/k_2 f(E^*) \qquad (7 - 24)$$

$$slope = 1/f(E^*)k_{-1} \qquad (7 - 25)$$

The unknowns yet to be determined are the effective number of oscillators and the activation energy, E^*. First, determine the activation energy by plotting $\ln\{k_2 f(E^*)\}$ versus $1/T$ using the intercept data; the slope will be the negative of the activation energy divided by the gas constant. The number of oscillators is defined by the relation

$$k_1 = k/[A] \qquad (7 - 26)$$

evaluated at the $[A]$ for which the observed rate constant begins to decrease. At this pressure it is assumed, by using Eq. (7 - 26), that the observed reaction rate is limited by the rate of activation. Then, the observed rate of reaction, $k[A]$, will be equal to the rate of activation by collisions, $k_1[A][A]$; hence the relation (7 - 26). In practice, that concentration of A for which the rate constant just begins to decrease is not easy to define. The following example problem will develop this point and describe two weaknesses of the Lindemann Theory.

Example 1 Isomerization of Methyl Isocyanide at 199.4°C   The observed first-order rate constants are given in Table 7 - 2 for the isomerization of methyl isocyanide to acetonitrile as a function of pressure. Values reported in columns two and three were multiplied by 10^7 and 10^3, respectively. The multipliers for the last two columns are 10^{-5} and 10^{-4}, respectively.

1. Find the rate constant at infinite pressure according to the Lindemann Theory and comment upon the linearity of the reciprocal rate constant plot.
2. Find the effective number of vibrators according to the Hinshelwood Theory if the activation energy is 160 kJ/mole and if the effective molecular diameter, d, is 4.5 angstroms. Compare this number to that calculated for the number of modes of vibration for this nonlinear molecule according to the 3N-6 rule.

Solution   The following results were obtained using the data of the ten highest pressures fit to a least-squares routine on the reciprocal relationship of $1/k$ versus $1/[A]$.

square of the corrrelation coefficient $= 0.9939$
slope $= 19.839$ M-s
intercept $= 13510$ s

Table 7 - 2 Unimolecular Rate Constants versus Total Pressure

p,torr	k, 1/s	[A], M	$1/k$, s	$1/[A]$, 1/M
0.161	1.01	0.00546	9.90	183.00
0.297	1.36	0.01008	7.35	99.20
0.648	2.83	0.02199	3.53	45.50
1.03	4.08	0.03495	2.45	28.61
2.03	5.91	0.06889	1.69	14.52
2.64	7.38	0.08959	1.36	11.16
5.00	11.7	0.16968	0.85	5.89
7.25	14.2	0.24604	0.70	4.06
10.0	18.0	0.33936	0.56	2.95
15.7	21.6	0.53279	0.46	1.88
24.8	30.	0.84161	0.333	1.188
35.7	34.	1.21151	0.294	0.825
72.3	45.	2.45357	0.222	0.408
101.5	52.	3.4445	0.192	0.290
155.2	55.6	5.2669	0.179	0.189
208	60.2	7.0587	0.166	0.142
570	69.8	19.343	0.143	0.0517
1143	73.0	38.788	0.137	0.0258
2248	73.7	76.288	0.136	0.0131
5010	74.9	170.019	0.135	0.00588
7106	74.9	241.149	0.134	0.00415

Reprinted with permission from F. W. Schneider and B. S. Rabinovitch, *J. Am. Chem. Soc.* **84**, (1962), p. 4215.   Copyright 1962, American Chemical Society.

The reciprocal of the intercept is the high pressure limit of the observed rate constant or

$$k(p \to \infty) = 0.00007402 \text{ s}^{-1}$$

Figures 7 - 2 and 7 - 3 show the reciprocal plots in the high and low pressure regions, respectively. Clearly, neither of these plots is strictly linear although portions of the curves may be fit by a linear, least-squares routine. One means to test the quantitative reliability of the Lindemann Theory is to calculate the value of the pressure or concentration necessary to reduce the rate constant to one-half its original value. The derivation begins by writing the rate constant for two cases: the high pressure limit and the intermediate pressure.

At high pressure,

$$k_\infty = k_2 f(E^*)[A]/\{(k_2/k_{-1}) + [A]\} = k_2 f(E^*)$$

and $f(E^*) = k_1/k_{-1}$; thus

$$k_\infty = k_2 k_1/k_{-1}$$

At moderate pressures,

$$k/k_\infty = ([k_2 f(E^*)[A]/\{(k_2/k_{-1}) + [A]\})/k_2 f(E^*)$$

and $k/k_\infty = 0.5 = [A]/\{(k_2/k_{-1}) + [A]\}$; thus

$$[A] = k_2/k_{-1} \text{ and } k_2/k_{-1} = k_\infty/k_1$$

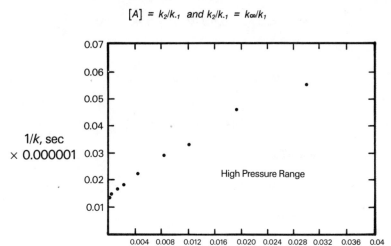

Figure 7 - 2 Reciprocal Rate Plot of Methyl Isocyanide (High Pressure)

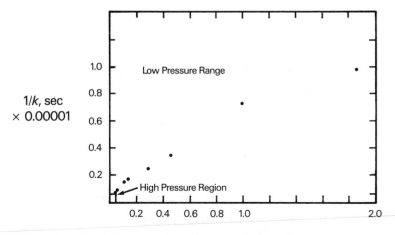

Figure 7 - 3 Reciprocal Rate Plot of Methyl Isocyanide (Low Pressure)

From the collision theory, $k_1 = 0.000404$ cc/mole-s using a diameter of 4.5 angstroms for the molecule. Thus, the predicted concentration of A at the half-value of k is 182.9 M; whereas the data show the proper [A] to be 1.36×10^{-3} M. Clearly, the Lindemann Theory cannot be used for any quantitative work.

To determine the effective number of vibrators, n, we evaluate $k/[A]$ as in the following.

$$k/[A] = 74.02 \times 10^{-6}/0.169 \; (M\text{-}s)^{-1} = 4.38 \times 10^{-4} M^{-1} s^{-1}$$

At the breakpoint,

$$\text{rate of activation } = \text{ rate of reaction}$$

$$k_1[A]^2 = k[A], \text{ so } k_1 = (k/[A])_{observed}$$

We use Eq. (7 - 21) to evaluate n.

$$k/[A] = 4.38 \times 10^{-4}/Ms = \{2\pi d^2/(n - 1)! \}(kT/\pi m)^{0.5}(E^*/RT)^{(n-1)} \exp(-E^*/RT)$$

$$E^*/RT = 160 \; kJ/mol/(8.314 \times 473) = 40.69$$

$$2\pi d^2 = 6.28(4.5 \times 10^{-10}m)^2 = 1.27 \times 10^{-18}m^2(6.02 \times 10^{23})$$

$$2\pi d^2 = 7.666 \times 10^5 m^2/mole$$

$$(kT/\pi m)^{0.5} = (8.314 \times 473/(3.14) \times 41/1000)^{0.5}$$

$$(kT/\pi m)^{0.5} = (3.055 \times 10^4)^{0.5} = 1.748 \times 10^2 m/s$$

so,

$$4.38 \times 10^{-4}/M\text{-}s = (7.66 \times 10^5 m^2/mol)(1.748 \times 10^2 m/s)^2(40.69)^{(n-1)}/(n - 1)!)(2.13 \times 10^{-18})$$

and with simplification,

$$1.537 \times 10^3 = (40.69)^{(n-1)}/(n - 1)!$$

By trial and error we find n to be between 3 and 4. This value of n differs from that calculated by the 3N-6 rule where the number of modes of vibration is 12.

This example problem illustrates two major faults of the Lindemann theory:
1. The plot of $1/k$ versus $1/[A]$ is not linear over the entire range. Portions of the curve may be considered linear for the purpose of curve-fitting, but the numerical results must be regarded with caution.

2. The limiting $[A]$ that is, when the collision rate limits the process predicted by the Lindemann Theory is much greater than what is observed. For the present case, $[A]$ at the half-value of k was predicted to be 182.9 M; whereas the data show this concentration at the half-value of k to be 1.36×10^{-3} M.

Furthermore, the number of vibrators is not equal to the number of modes predicted by the 3N-6 rule. It was hoped that some relation could be established between the parameter, n, and a characteristic of the activated complex.

Rice, Ramsperger, and Kassel Theory (RRK)

The Hinshelwood Theory suffers from unrealistic assumptions that the required activation energy need not be localized in a critical bond and that energies in excess of the minimum are not employed to increase the probability of the unimolecular reaction. The modified theory of Rice and Ramsperger[7, 8, 9] assumes the following:

1. The minimum activation energy must be in one vibrational bond for decomposition to occur.
2. Energies in excess of the minimum enhance the probability of reaction.

The probability of bond dissociation among n bonds may be derived as a function of E^* and n as follows.

Consider a molecule which is an ensemble of n oscillators coupled in such a way that energy E is exchanged within a molecule at frequency, w. This assumption pertains to the decomposition rate constant, k_2. We further assume that the redistribution of energy *within* the molecule occurs randomly. Statistical methods can be used to determine this energy reallocation within the reacting molecule. The molecule of n oscillators will decompose when one has acquired m quanta or greater out of a total of j quanta. Thus, we calculate first the number of ways, $q(n, j)$, of distributing the j quanta among the n oscillators.

$$q(n, j) = [(j + n - 1)!]/[j!(n - 1)!]$$

Now for the requirement that m quanta be localized in one bond before the decomposition occurs; then the number of ways, $q_m(n, j)$, of distributing these quanta among n oscillators is

$$q_m(n, j) = [(j - m + n - 1)!]/[(j - m)!(n - 1)!]$$

The probability, P, of this critical configuration occurring is the ratio of q_m/q or

$$P = [(j - m + n - 1)!j!]/[(j - m)!(j + n - 1)!]$$

We may factor these polynominals as follows:

$$P = \{(j - m + n - 1)!/(j + n - 1)(j + n - 2)...$$
$$(j + n - m - 1)!\}\{j(j - 1)...(j - m)!/(j - m)!\}$$

$$P = (j + n - 1 - m)(j + n - 2 - m)...$$
$$(j + n - n - m)!/(j - m)!(j + n - 1)(j + n - 2)...(j - 1)$$

$$P = \Pi(j - m + s)/(j + s), \text{ where } s = (n - 1)$$

Now divide within the extended product:

$$
\begin{array}{r}
1 - m/(j + s) \\
\hline
(j + s) \ \overline{\big)\ j - m + s} \\
j + s \\
\hline
-m \\
-m \\
\hline
0
\end{array}
$$

Thus, $P = \Pi[1 - m/(j + s)]$; and assume $j \gg (n - 1)$ so $j + s = j$ for all s; then

$$P = [1 - m/j]^{(n-1)} = [1 - E^*/E_j]^{(n-1)} = [(E_j - E^*)/E_j]^{(n-1)}$$

where E_j and E^* are the energies of the jth and mth quantum states, respectively. The decomposition rate constant is the product of the probability with the frequency of the energy exchange, w.

$$k_2 = wP = w[(E - E^*)/E]^{(n-1)} \tag{7 - 27}$$

It is assumed that this molecule is in thermal equilibrium with a bath of M molecules at temperature T. Extramolecular energy exchange by collison "fuels" the decomposition reaction. Statistics gives for such a collection of molecules (one having n oscillators requiring minimum energy, E^*) the energy dependent probability that one such oscillator has been activated as follows:

$$f(E)\, dE = E^{(n-1)} \exp(-E/kT)/[(n - 1)!(kT)^n]\, dE \tag{7 - 28}$$

The infinite pressure limit of k is given by

$$k_\infty = f(E^*)k_2$$

or

$$k_\infty = \int_{E^*}^{\infty} E^{(n-1)} \exp(-E/kT)/[(n - 1)!(kT)^n]w\{(E - E^*)/E\}^{(n-1)}\, dE$$

Let $u = (E - E^*)/kT$ and $du = dE/kT$; then

$$k_\infty = w \exp(-E^*/kT)/(n - 1)! \int_0^{\infty} u^{(n-1)} \exp(-u)\, du$$

$$k_\infty = w \exp(-E^*/kT) \tag{7 - 29}$$

For lower pressures we return to Eq. (7 - 22)

$$k = f(E^*)k_{-1}[M]k_2/\{k_2 + k_{-1}[M]\}; \quad k_{-1} = \text{reverse rate constant}$$

then substituting for $f(E^*)$ and k_2 yields

$$k = \int_{E^*}^{\infty} \{k_{-1}[M]E^{(n-1)} \exp[-E/kT]/[(n-1)!(kT)^n]w[(E-E^*)/E]^{(n-1)} dE/$$
$$\{w[(E-E^*)/E]^{(n-1)} + k_{-1}[M]\}\} \qquad (7 - 30)$$

As before, let $u = (E - E^*)/kT$, divide numerator and denominator by $k_{-1}[M]$, and let $k_\infty = w \exp(-E^*/kT)$:

$$k/k_\infty = 1/(n-1)! \int_0^{\infty} u^{(n-1)} \exp(-u) du/\{1 + w/k_{-1}[M]\{u/(u + E^*/kT)\}^{(n-1)}\} \qquad (7 - 31)$$

E^*/kT and w are known from the Arrhenius plot of $\ln(k_\infty)$ versus $1/T$. The parameter n is determined from data of k versus $[M]$.

RRKM

The Marcus modification of the RRK theory allows one to calculate $k_2(E)$ directly from a consideration of the reacting molecule structure. The theory of intramolecular energy redistribution is refined to include the concept of random lifetimes. Thus, the generation rate of A^* depends upon the energy distribution within the molecule and the lifetime of the same. We write the differential rate of production as

$$dR(e,t) = [\text{rate of generating } A^* \text{ molecules by collisions}] \times$$
$$[\text{probability of having lifetimes } t \text{ and } t + dt] \times$$
$$[\text{probability of reacting before collision}]$$

The first term in this equation is given by the product of the collision frequency, Z, the concentration of reacting and inert molecules, and the probability of forming an A molecule having energy between E and $E + dE$, or

$$[\text{rate of gen. } A^*] = Z[A][M]P(E) dE \qquad (7 - 32)$$

The distribution function, $P(E)$, will be discussed in more detail later. Now, to the problem of defining the lifetime, t. This is the interval between the time of collision and the time which the products are formed. The probability of forming molecules with the required energy in the lifetime t to $(t + dt)$ is just the fraction of initial number of molecules remaining energized at $(t + dt)$ minus the fraction of initial number of the molecules remaining energized at t.

$$P(t,E) \, dE = \{[\text{fraction of A molecules remaining energized}]_{(t+dt)} -$$
$$[\text{fraction of A molecules remaining energized}]_t\} dt$$

$$P(t,E) \, dE = -d/dt\{[A^*]/[A^*]_o\} \, dt$$

The negative sign before the derivative is necessary so that the probability remains positive. Now for the unimolecular decomposition *alone*, the rate is proportional to the $[A^*]$ where the proportionality constant, k_2, is a function of energy, E.

$$-d[A^*]/dt = k_2(E)[A^*]$$

for which upon integration between zero time and t gives

$$ln([A^*]/[A^*]_o) = -k_2(E)t$$

and

$$[A^*]/[A^*]_o = exp(-k_2(E)t)$$

for which

$$-d/dt([A^*]/[A^*]_o) = k_2(E) \, exp(-k_2(E)t)$$

The final expression for the probability of A^* molecules having lifetimes between t and $(t + dt)$ is

$$P(t,E) = k_2(E) \, exp[-k_2(E)t]$$

It is assumed that the frequency of deactivation is given by $w_d = Z[M]$. The probability of reaction, $P_{rxn}(t)$, is just the probability of A^* molecule not deactivated within the lifetime. The probability of not being deactivated by collision is

$$w P_{reaction}(t) = w_d P_{not\ deactivated}$$
$$= Z[M](\text{fraction of molecules not deactivated at } t)$$
$$= Z[M]([A^*]/[A^*]_o)$$

Now the rate of deactivation *alone* is given by

$$r_d = -d[A^*]/dt = Z[M][A^*]$$

or

$$[A^*]/[A^*]_o = exp(-Z[M]t)$$

thus

$$P_{reaction}(t) = exp(-Z[M]t)$$

The differential expression for rate can be integrated as follows. First integrate on t from zero to infinity

$$dR\ dt = Z[M][A]P(E)P(t,E)P_{rxn}(t)\ dE\ dt$$

$$\int_0^\infty dR\ dt = Z[M][A]P(E)\ dE\ k_2(E)\int_0^\infty exp(-k_2(E)t)\ exp(-Z[M]t)\ dt$$

$$dR(E) = Z[M][A]P(E)k_2(E)\ dE/\{k_2(E) + Z[M]\} \qquad (7 - 33)$$

Next divide by $[A]$ and multiply numerator and denominator by $1/(Z[M])$. Division of $dR(E)$ by $[A]$ yields $dk'(E)$ for linear kinetics

$$dk'(E) = k_2(E)P(E)dE/\{1 + k_2(E)/(Z[M])\}$$

Integrate on E between E^* and infinity:

$$\int_0^\infty dk'(E) = \int k_2(E)P(E)dE/\{1 + k_2(E)/(Z[M])\} \qquad (7 - 34)$$

Reevaluate $P(E)$ using a quantum statistical formula for probability of any system with quantum density $g(E)$ in the range, $E + dE$,

$$P(E)\ dE = [g(E)\ exp(-E/kT)/q_v]\ dE$$

where q_v is the vibrational partition function. In the integration the lower limit is E^* thus we need the g for energized molecules only, $g'(E)$. Substituting this function for $P(E)$ into Eq. (7 - 34) gives

$$k' = (1/q_v)\int_{E^*}^\infty k_2(E)g'(E)\ exp(-E/kT)\ dE/\{1 + k_2(E)/(Z[M])\} \qquad (7 - 35)$$

Next we turn our attention to the description of $k_2(E)$. Consider the distinction between those molecules in the critical configuration, A^c, and those not in the critical configuration, A^*. These two are different by virtue of the distribution of energy within the molecule. The A^c molecule is critical because the addition of a small amount of energy in one bond will result in rupture of that bond. Thus, we have the equilibrium

$$A^* \Leftrightarrow A^c$$

The ratio of concentrations, $[A^c]/[A^*]$, is given by the ratio of the quantum densities.

$$[A^c]/[A^*] = g^c(E)/g^*(E)$$

Since an equilibrium exists between the two then

$$k_2(E)[A^*]dE = k_2{}^c(E)[A^c]dE$$

and

$$k_2(E) = k_2{}^c(E)[A^c]/[A^*] = k_2{}^c(E)g^c(E)/g^*(E)$$

The random lifetime assumption states this equation is true at all pressures. We express $k_2{}^c(E)$ in terms of parameters pertaining to the molecular description. We assume the critical coordinate behaves as translation over a small distance x. Only a part of the energy E of an A^c molecule will be translated along the critical coordinates. Call this e_t which differs from one A^c to another

$$e_t = 0.5mv^2 \qquad (7 - 36)$$

where

m = effective mass

v = velocity along critical coordinate

The maximum value of e_t is $E - E^*$. The rate constant for crossing one-half the length, x, is

$$k_2{}^c = v/2x = (e_t/2m)^{0.5}/x \qquad (7 - 37)$$

An alternate expression for x comes from the particle is a box problem of quantum mechanics.[10] Here the eigen energy, e_t, is given by

$$e_t = (hn/x)^2/8m$$

where n equals the quantum number in the x direction. The density of states, g^c, is just the energy derivative of the quantum number, n.

$$g^c(e_t) = dn/d(e_t) = 0.5(8me_t x^2 h^2)^{-0.5}(8mx^2/h^2) \qquad (7 - 38)$$

$$g^c(e_t) = (x/h)(2m/e_t)^{0.5}$$

The product of the rate constant, $k_2{}^c(E)$, and the density of states for the molecule in the critical configuration is equal to the integral over all energies, e_t, between 0 and $E - E^*$ for the product of the rate constant, $k_2{}^c(e_t)$ and the densities of quantum states, $g^c(e_t)$ and $g^c(E - e_t)$

$$k_2{}^c(E)g^c(E) = \int k_2{}^c(e_t)g^c(e_t)g^c(E - e_t)d(e_t)$$

where

$k_2{}^c(E)$ = average rate constant for the molecules in critical configuration with total energy, E

$g^c(E)$ = density of quantum states in critical configuration between energies, E and $E + dE$

$k_2{}^c(e_t)$ = rate constant for molecule in critical configuration needing energy, e_t, to react along critical coordinate

$g^c(e_t)$ = density of quantum states in critical configuration along the critical coordinate having energy, e_t

$g^c(E - e_t)$ = density of quantum state in critical configuration along all other coordinates (i.e., all other nonreactive vibrations) at energy, $E - e_t$

The product of $g^c(e_t)g^c(E - e_t)$ gives the number of ways that the total energy, E, is distributed among all the possible reaction coordinates in the molecule requiring energy, e_t. When this degeneracy is multiplied by the rate constant along one critical coordinate having energy, e_t, $k_2{}^c(e_t)$, and integrated over all possible values of e_t ($e_t \rightarrow 0$ to $E - E^*$, since E^* is the minimum energy to put into the critical state), then the average rate constant times the degeneracy of the entire molecule is the result (h is Planck's constant).

$$k_2{}^c(E)g^c(E) = 1/h \int g^c(E - e_t) \, de_t$$

Consider now expressions for $g^c(E)$ and so on to allow calculations of $k(T)$ by the RRKM method. The working expression is

$$k_2(E) = (1/hg^*(E)) \int g^c(E - e_t) \, de_t = N_c(E - E^*)/hg^*(E)$$

Next we must have formulas for Z, q_v, $g^*(E)$, and $N_c(E - E^*)$. Begin with Z in the units of cm^3/sec:

$$Z = d^2(8\pi kT/\mu)^{0.5}$$

d = collision cross sectional diameter

k = Boltzmann constant

T = absolute temperature, $^\circ$K

μ = reduced mass

For inert molecule, M, the factor, $Z[M]$, is conveniently written as a function of partial pressure of M as

$$Z[M] = (4.41 \times 10^7) d^2 T^{-0.5} \mu^{-0.5} p$$

where

$$Z[M] \quad [=] \sec^{-1}$$
$$d \quad [=] \text{angstroms}$$
$$p \quad [=] \text{Torr}$$

The partition function, q_v, is evaluated by standard procedures discussed in the transition state theory and will not be presented here. The formulas for $g^*(E)$ and $N_c(E - E^*)$ can be derived by formulas shown for the RRK theory but corrected for zero-point energies of the vibrator.

$$g^*(E) = (E + E_z^*)^{(n-1)}/(n - 1)! \prod_{i=1}^{n} h\nu_i^* \tag{7 - 39}$$

$$N_c(E - E^*) = (E - E^* + E_z^c)^{(n-1)}/(n - 1)! \prod_{i=1}^{n-1} h\nu_i^c \tag{7 - 40}$$

where $E_z^* = (1/2)\sum h\nu_i^*$ and $E_z^c = (1/2)\sum h\nu_i^c$ are zero-point energies and ν_i^c, ν_i^* are fundamental frequencies of A^c and A^*, respectively. Thus, we have the following for $k_2(E)$:

$$k_2(E) = N_c(E - E^*)/hg^*(E) = \{(E - E^* + E_z^c)/(E + E_z^*)\} \prod_{i=1}^{n} \nu_i^* / \prod_{i=1}^{n-1} \nu_i^c \, (q_r^c/q_r^*)$$

$$(q_r^c/q_r^*) = \{[I_x^c \, I_y^c \, I_z^c]/ I_x^* \, I_y^* \, I_z^*]\}^{1/2}$$

Homogeneous Bimolecular Reactions

The presentation of bimolecular homogeneous reaction theory will be brief owing to the lengthy discussion of unimolecular decomposition reactions in which we described many of the fundamentals of bimolecular reactions. The discussion here will focus on the system of atom A and molecule BC to form molecule AB and atom C in an irreversible reaction. The rate of reaction is given by

$$R = [A][BC] \int (v)(Q[v])(f[T,v]) \, dv \tag{7 - 41}$$

where

v = relative velocity of the atoms, molecules
T = temperature, $^{\circ}K$
$Q[v]$ = reactive cross sectional area, a function of v
$f[T, v]$ = distribution function at velocity, v, and temperature, T; gives the number of atoms/molecules having velocity between v and $v + dv$

$[A]$,
$[BC]$ = molar concentrations of A, BC

The reader is referred to Reference 11 for a derivation of Eq. (7 - 41).

Reactive Cross Sectional Area

The reactive cross sectional area depends upon the relative velocity of the system; however, several important features are concealed in this simple statement. The energy states of the individual atoms and molecules are important to the reactive cross section. That is, the electronic, translational, rotational, and vibrational states of the reactants and products must be considered before Q can be specified. The reactive process requires energy transfer through a collision; however, the effectiveness of any one collision depends upon the "trajectories" of the collision, before and after. Trajectory here means not only the speed and direction of the species but also the orientation of the species and the energy distribution within the bonds of the reactant molecule. The desired product, AB, will be most likely to occur for only certain orientations of the BC reactant molecule with respect to the A atom. Thus, collisions between the reactants having unfavorable orientations will make only small contributions to the total observed rate. By a similar argument, certain energy distributions within the BC molecule prior to collision with A will render it more reactive than other distributions even though in both cases the BC molecule may contain the same total energy. Thus, we may reformulate Eq. (7 - 41) to include these considerations, and using Maxwell-Boltzmann statistics to specify the velocity distributions of the atoms/molecules gives

$$R = [A][BC]\{8/[\mu\pi(kT)^3]\}^{0.5} \int E\ Q(E,T)\ exp(-E/kT)\ dE \qquad (7 - 42)$$

where

μ = reduced mass: $m_1 m_2/(m_1 + m_2)$

m_1, m_2 = mass of atom A, molecule BC, respectively

k = Boltzmann constant, 1.38×10^{-16} erg/mole-K

T = absolute temperature, $^{\circ}$K

$Q(E,T)$ = reactive cross sectional area

E = energy of system

For Eq. (7 - 42) we have isolated all of our ignorance of the reactive process into one function: $Q(E,T)$. The task of the theoretician is to propose physical models of the process and solve the mathematics to yield $Q(E,T)$.

Collision Theory for Bimolecular Reactions

The most naive model for $Q(E,T)$ assumes that a step function describes the energy dependence and the hard-sphere collison diameter, d_c, describes the reactive cross sectional area. Thus, the function for $Q(E,T)$ by this model is given by

$$Q(E,T) = \pi(d_c)^2 S(E - E^*) \qquad (7 - 43)$$

where E^* is the threshold energy for reaction, $S(E - E^*)$ is the unit step function, and d_c is the sum of reactant atom/molecule radii. Upon substituting Eq. (7 - 43) into Eq. (7 - 42) we have

$$k(T) = R/[A][BC] = (8\pi kT/\mu)^{0.5}(d_c)^2 \exp(-E^*/kT)(1 + E^*/kT) \qquad (7 - 44)$$

Equation (7 - 44) suffers from one fault: the rate constant does not account for the angle of incidence at the collision. By this model, collisions of grazing incidence are just as effective as head-on collisions. This problem may be corrected using the following expression for $Q(E,T)$.

$$Q(E,T) = \pi(d_c)^2[1 - (E^*/E)]S(E - E^*) \qquad (7 - 45)$$

When this expression is substituted into Eq. (7 - 42) the collision theory rate constant is the result.

$$k(T) = (8\pi kT/\mu)^{0.5}(d_c)^2 \exp(-E^*/kT) \qquad (7 - 46)$$

Equation (7 - 46) is also known as the hard-sphere collision frequency. When the rate constants predicted by Eq. (7 - 46) are divided into the observed rate constant, this ratio is designated as p. Values of p are observed to vary between 0.001 and unity.[11] Only for bimolecular reactions involving alkali metal atoms and certain charged species is p greater than unity. These data suggest that the assumptions inherent to the development of the reactive cross sectional area overpredict the efficiency of the collisions. The unit, step function as a threshold is probably unrealistic and refinements to the theory show a better function to model $Q(E,T)$. Some have avoided the problem by adding a fitting parameter, namely p, to Eq. (7 - 46) which reduces it to a correlating equation.

Refinements to the Collision Theory

Studies of reactions in crossed molecular beams have shed light on the general behavior of reactive cross sections in bimolecular reactions: $Q(E,T)$ rises sharply at the threshold energy, but decreases after attaining a maximum value. A number of mathematical equations are available to model this behavior; but one particularly useful equation is

$$Q(E,T) = a(E - E^*)^n \exp[-b(E - E^*)]S(E - E^*) \qquad (7 - 47)$$

where $S(E - E^*)$ is the unit, step function, a and b are adjustable parameters. This function, when manipulated according to Eq. (7 - 42), gives the rate constant.

$$k(T) = [8/\pi\mu(kT)^3]^{0.5}[a\Gamma(n + 1)/c^{(n+1)}][(n+1)/c + E^*][exp(-E^*/kT)] \qquad \text{(7 - 48)}$$

$$c = b + 1/kT$$

$\Gamma(n + 1)$ = gamma function = n! for positive integers

However, without some prior knowledge of a, b, and n; the usefulness of Eq. (7 - 48) is limited to curve-fitting of experimental data.

Homogeneous Termolecular Reactions

Recombination reactions (such as A + A $\Leftrightarrow$ A$_2$) carried out in high dilution with an inert, M, show rate laws of third order, overall. The necessity of the inert, or chaperon, atom/molecule was recognized from a consideration of the potential energy curves for recombination reactions. Consider the potential curve (Fig. 7 - 4) for two H atoms approaching to form a nonrotating, hydrogen molecule ($J = 0$). As the interatomic distance decreases, the potential for the system decreases thus accelerating the pair towards one another. Unless some of the initial kinetic energy is removed from the system, the atom pair will continue to approach one another causing the potential energy to rise until it reverses the process with the atoms speeding *away* from one another at a kinetic energy equal to the precollision state. Thus, the role of the chaperon species is to remove that energy necessary to stabilize the atom pair at a potential energy less than the combined kinetic energies before collision. This three-body process may be modeled as a single mechanistic step.

$$H + H + Ar \rightarrow H_2 + Ar \qquad \text{(7 - 49)}$$

Another interpretation of these termolecular rate laws would have the formation of an unstable complex involving the chaperon species as part of the reaction mechanism. Consider the recombination of iodine atoms with the chaperon, M.

$$I + M \Leftrightarrow IM^* \qquad \text{(7 - 50)}$$

$$IM^* + I \rightarrow I_2 + M \qquad \text{(7 - 51)}$$

Data for the iodine recombination have been successfully correlated by this mechanism for a number of different chaperon species (Fig. 7 - 5). Surprisingly, some chaperons result in data having negative activation energies; that is, increasing temperatures show rate constants that decrease in value.

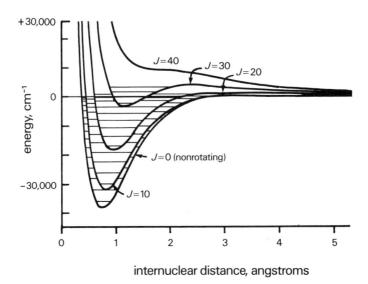

internuclear distance, angstroms

Figure 7 - 4   Potential Energy Curves for Molecular Hydrogen

Potential energy curves for the hydrogen molecule in rotational states J = 0, 10, 20, 30, and 40. These rotational states have 15, 13, 9, and 4 vibrational states, respectively. The energies of the vibrational states are indicated by horizontal lines. Rotational states with J = 35, 36, and 37 have one vibrational state, and there are no bound states for J > 37. Potential energy curves for other diatomic molecules are similar to these except that the numbers of rotation and vibration states are larger. Potential energy curves to which rotational kinetic energy has been added are frequently called effective potential energy curves. Used with permission from W. C. Gardiner, *Rates and Mechanisms of Chemical Reactions* (New York: W. A. Benjamin Inc., 1969).

Transition State Theory (TST)

Potential energy curves are familiar and useful tools in discussing the interactions between two species. The transition state theory (also known as the Activated Complex Theory) employs potential surfaces to describe the interactions between three species as in the reaction

$$A + BC \Leftrightarrow AB + C \qquad (7 - 52)$$

The discussion begins with a description of the proper potential energy surface to model Eq. (7 - 52), followed by the assumptions of the transition state theory.

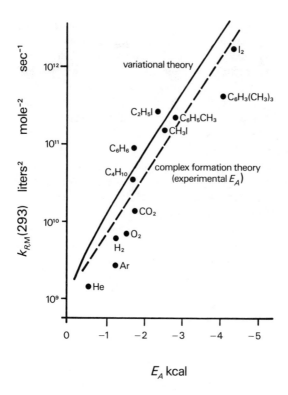

Figure 7 - 5 Comparison of Predicted and Experimental Rate Constants for Iodine Recombination Reaction

Used with permission from W. C. Gardiner, *Rates and Mechanisms of Chemical Reactions* (New York: W. A. Benjamin, Inc., 1969).

Potential Energy Surfaces

The motions of atoms A, B, and C would normally require that nine coordinates be specified for which the forces on each may be derived from Newtonian mechanics. The force in the *x*-direction of the *ith* atom is given by the partial derivative of the potential energy surface given by $V(x,y,z)$

$$f_{i,x} = (\partial V/\partial x_i) \tag{7 - 53}$$

Once again, it would seem that nine such partial derivatives would be necessary to describe the forces on A, B, and C; however, a little reflection shows that only the relative position vectors are necessary for a nonrotating system. For rotating systems, the potential energy may be altered by changing the rotational velocity even though the relative positions may remain fixed. Thus, for $J = 0$, the relative position vectors, r_{AB}, r_{BC}, and r_{AC} connecting the atoms A, B, and C are sufficient to describe the system. For a rotating system, V would also depend upon the components of rotational velocity: a_x, a_y, and a_z. Accurate calculations must include the rotational states; however, this development shows that the rotational and relative position vectors (i.e., r_{AB}, and so on) are separable.

Consider now a nonrotating system for which only the relative position vectors are required to specify the potential surface. The reacting system is endothermic for which the activation energy is greater than the enthalpy of reaction, ΔH. Suppose the initial position shows atoms B and C bonded together to form molecule BC; whereas A is an isolated atom. The vector, r_{BC}, is small near that of the internuclear distance; whereas the vectors, r_{AC} and r_{AB}, are large. If we define this configuration as having zero potential energy, then reducing the separation vectors, r_{AC} and r_{AB}, will cause the total potential energy to increase due to repulsion effects of the overlapping, electronic clouds. As the activation energy barrier, E_b, is surpassed, the reaction may occur to produce the AB molecule and the atom C. After this reaction event, the species separate, thus causing V to decrease to a new "equilibrium" position ΔH units above the initial position.

Some reactions of the form A + BC have been examined by semiempirical quantum mechanical calculations (e.g., extended Huckel MO and others) to show the collinear encounter is the most favorable one (i.e., has the lowest barrier energy). Now $V(x,y,z)$ becomes only a function of two separation vectors, r_{AB} and r_{BC}. This function can be represented in three-dimensional space. Consider the reaction of

$$O + H_2 \rightarrow OH + H$$

and the corresponding potential energy surface (Fig. 7 - 6). Several features are worthy of comment. Three reaction channels have been designated: alpha, beta, and gamma which lead to these products: O + H_2, OH + H, and O + H + H, respectively. The dissociation energies, D_o, for H_2 and OH are given as the heights above the zero-point energies on each ordinate. The activation energies for the forward and reverse reactions are given as E_{bf} and E_{br}, respectively. This surface presents several pathways, or trajectories, leading from the reactants to the products. Depending upon the initial states of the reactants, one could imagine an optimum trajectory for the collinear reaction:

$$A + BC \rightarrow AB + C$$

In the transition state theory, it is assumed that one minimum energy trajectory exists and this trajectory is designated the "reaction coordinate." This reaction coordinate

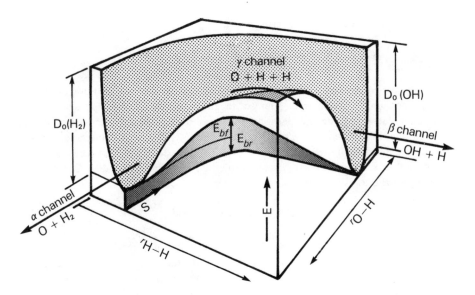

Figure 7 - 6 Potential Energy Surface for Collinear Reactive Encounter O + H$_2$

Experimental data for the H$_2$ + OH molecular potential curves were obtained in accurate spectroscopic experiments. The activation energy for this reaction is about 10 kcal, from which a barrier height for the forward reaction E$_{bf}$ of about 10 kcal = 3,500 cm^{-1} can be estimated. The heat of reaction is -1.9 kcal = -670 cm^{-1}. Reaction into the gamma channel to give three atoms would not be observed experimentally owing to the high energy necessary. Reactive encounters originating from the gamma channel are, in principle, observable, but would be complicated by a fourth channel (not shown here) corresponding to an excited electronic state of OH with energy 32,720 cm^{-1} greater than the ground electronic state. Reactions that occur on a single potential surface are called adiabatic; reactions described by crossings between surfaces are called nonadiabatic reactions. Used with permission from W. C. Gardiner, *Rates and Mechanisms of Chemical Reactions* (New York: W. A. Benjamin, Inc., 1969).

defines a specific molecular motion which *leads directly to the reactive event*. In the transition state theory this reaction coordinate is separable from all other motions of the reactant species.

The formalism of the TST, constructed for the bimolecular reaction

$$A + B \rightarrow products$$

will be developed by first postulating an equilibrium exists between the reactants and the transition state, C, as follows:

$$A + B \Leftrightarrow C \rightarrow products \qquad (7 - 54)$$

The "equilibrium" constant is described by

$$K^* = [C]/[A][B] = \{q_c^*/[q_a^* q_b^*]\} \exp(-E_b'/kT)$$

where the concentrations are expressed in molecules/cm^3 and the partition functions, q_i^*, are per unit volume and referred to their individual zero of energy. The Boltzmann factor adjusts the partition functions to a common zero-energy datum through the barrier energy, E'_b. Refer to Figure 7 - 7 for an illustration of the terms. The partition functions describe each of the external and internal modes of energy and as such each function is separable into the different modes. Consider the function for the transition state, C.

$$q_c^* = q_c(external)q_c(internal) \qquad (7 - 55)$$

The external mode is translation; whereas the internal modes are electronic, rotational, and all vibrational modes other than the reaction coordinate. The reaction coordinate is considered one of translation rather than vibration because no periodic motion is observed in this critical reaction coordinate; the bond is

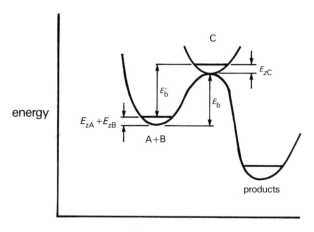

reaction coordinate

Figure 7 - 7   Schematic of Potential Energy Curve Describing the TST

Schematic diagram indicating the accounting procedure for zero-point energy in activated complex calculations. The barrier height E_b is the same quantity shown in Figure 3 - 1. The energy E'_b is the barrier height corrected for the zero-point energies of the reactants and the activated complex by the formula $E'_b = E_b + E_{zC} - E_{zA} - E_{zB}$. Each of the zero-point energies E_z is given by a sum over the appropriate set of vibrational frequencies $E_z = (1/2)\sum h\nu_i$. Used with permission from W. C. Gardiner, *Rates and Mechanisms of Chemical Reactions* (New York: W. A. Benjamin, Inc., 1969).

Chapter 7   Homogeneous Kinetics

completely activated upon the first tranverse of this internal vibration mode. As a translation the partition function for the reaction coordinate may be written as

$$q_c(reaction\ coordinate) = (2\pi\mu kT)^{0.5}(d/h) \qquad (7 - 56)$$

where d is the length of the critical coordinate, h is Planck's constant, k is Boltzmann's constant, and μ is the reduced mass of the complex, C. Thus, we write the partition function for the complex as

$$q_c^* = (2\pi\mu kT)^{0.5}(d/h)q_c' \qquad (7 - 57)$$

The partition function, q_c', is that for the remainder of the complex, internal and external, other than the critical coordinate. Using this expression for the partition function of the complex and the equilibrium relationship, one may solve the concentration of complexes.

$$[C] = (2\pi\mu kT)^{0.5}(d/h)(q_c'/[q_a^*q_b^*])\ exp(-E_b'/kT)[A][B] \qquad (7 - 58)$$

The density of activated complexes per unit length of reaction coordinate is just the derivative of Eq. (7 - 58) with respect to d, or

$$d[C/]dd = (2\pi\mu kT)^{0.5}(1/h)(q_c'/[q_a^*q_b^*])\ exp(-E_b'/kT)[A][B] \qquad (7 - 59)$$

At the dynamic equilibrium postulated by the TST, half of the complexes are moving to one side of the equilibrium according to an average velocity given by the kinetic theory of gases. This velocity is

$$v = (2kT/\pi\mu)^{0.5}$$

The rate of complex formation is Eq. (7 - 59) divided by two and multiplied by the velocity. The reaction rate constant for the bimolecular reaction is given by the following:

$$k(T) = (2\pi\mu kT)^{0.5}(2kT/\pi\mu)^{0.5}(1/2h)(q_c'/[q_a^*q_b^*])\ exp(-E_b'/kT) \qquad (7 - 60)$$

Simplification of the equation leads to

$$k(T) = (kT/h)(q_c'/[q_a^*q_b^*])\ exp(-E_b'/kT) \qquad (7 - 61)$$

Conclusions

The discussion presented in this chapter concerning homogeneous reactions shows a unifying feature: inert species play a significant role in the reaction mechanisms. In two cases, the inert atom/molecule served to bring about the reaction either by supplying the energy of activation (unimolecular decomposition) or removing energy (recombination reactions). Energy transfer was necessary to produce the desired reaction condition. The need for a reactive chaperon in these homogeneous reactions suggests the same need may be present in the heterogeneous reactions. Thus, the catalyst may be just an immobile but multipurpose chaperon to perform the tasks of energy transfer, reactant orientation (via selective chemisorption), and stabilization of reactive intermediates.

REFERENCES

1. Perrin, M. W., *Ann. Phys. Paris*, **11**, (1919), p. 5.

2. Daniel, F. and E. H. Johnson, *J. Am. Chem. Soc.*, **43**, (1921), p. 73.

3. Lindemann, F. A., *Trans. Faraday Soc.*, **17**, (1922), p. 598.

4. Schneider, F. W. and B. S. Rabinovitch, *J. Am. Chem. Soc.*, **84**, (1962), p. 4215.

5. Chan, S. C., B. S. Rabinovitch, L. D. Spicer, T. Fugimoto, Y. N., Lin, and S. P. Pavlou, *J. Phys. Chem.*, **74**, (1962), p. 3160.

6. Hinshelwood, C. N., *Proc. Roy. Soc. (London)*, **113A**, (1927), p. 230.

7. Rice, O. K. and H. C. Ramsperger, *J. Am. Chem. Soc.*, **49**, (1927), p. 1617; **50**, (1928), p. 612; O. K. Rice, *Proc. Nat'l Acad. Sci., USA*, **14**, (1928), pp. 114, 118.

8. Kassel, L. S., *J. Phys. Chem.*, **32**, (1928), pp. 225, 2065.

9. Marcus, R. A., *Chemische Elementarprozess*, Ed. by H. Hartman (Berlin: Springer Verlag, 1968).

10. Hanna, M. W., *Quantum Mechanics in Chemistry*, Second Edition (New York: W. A. Benjamin, Inc., 1969).

11. Gardiner, W. C., *Rates and Mechanisms of Chemical Reactions* (New York: W. A. Benjamin, Inc., 1969).

12. Bodenstein, M., *Z. Physik Chem.*, **100**, (1922), p. 106.

PROBLEMS

1. The unimolecular decomposition of azomethane, $CH_3-N=N-CH_3$, to ethane and nitrogen has been studied. The "high" pressure rate constant is given by

$$k_\infty = 3.13 \times 10^{16} \, exp(-52,440/RT) \qquad (7 - 62)$$

where the units of the rate constant is reciprocal seconds. Experimental values of the rate constant at $603°K$ and for several pressures are given in the following table.

Chapter 7 Homogeneous Kinetics

Using the RRK theory, calculate the rate constant at the following pressures: 1, 10, 60, 100 Torr. Present your data as a plot of $\log_{10}(k/k_\infty)$ versus p. Assume the effective number of oscillators is 24 and the hard-sphere diameter is 13 angstroms. Put the experimental points on the graph.

2. As an alternative to the derivation of k_∞ equation in the RRK theory, show that s-fold repeated applications of the differentiation by parts will lead from the following equation

$$k_\infty = \int w(E - E^*)^{(s-1)} exp(-E/kT) \, \Gamma(s)^{-1} \, (kT)^{-s} \, dE \qquad (7 - 63)$$

to the following equation for positive, integral values of s.

$$k_\infty = w \, exp(-E^*/kT) \qquad (7 - 64)$$

Used with permission from W. C. Gardiner, *Rates and Mechanisms of Chemical Reactions* (New York: W. A. Benjamin, Inc., 1969).

3. Calculate the ratios k_1/k, at 1000 K for a reaction with $E_A = 40$ kcal/mol and for $n = 1$, 4, 12, and 20 using the Hinshelwood modification to the Lindemann Theory (Eq. 7 - 17). Calculate the pressure when $k = 0.5 \, k_\infty$ for $n = 1$, 4, 12, & 20 if $k_\infty = 10^7$ /sec. Used with permission from W. C. Gardiner, *Rates and Mechanisms of Chemical Reactions* (New York: W. A. Benjamin, Inc., 1969).

4. A linear approximation is sometimes used for describing atom-atom collisions near the threshold energy, E^*. Show that the function $Q(E) = a(E - E^*)$ leads to the rate constant formula

$$k(T) = a[8(kT)^3/\pi\mu]^{1/2} \, (E^*/kT + 2) \, exp(-E^*/kT) \qquad (7 - 65)$$

The cross section for exciting an H atom to its first electronically excited state (principal quantum number = 2) by a collision with another H atom was found to be

$Q(E) = 1.3 \times 10^{-18} (E - 10.2)$ cm^2 where E is the energy in eV and 10.2 is the threshold energy for excitation, also in eV. Calculate the rate constant for this process at $T = 12,000$ K. Used with permission from W. C. Gardiner, *Rates and Mechanisms of Chemical Reactions* (New York: W. A. Benjamin, Inc., 1969).

5. The decomposition of nitrogen dioxide to nitric oxide and oxygen is second order in the reactant. Bodenstein[12] reported the rate constant for this reaction as a function of temperature (Table 7 - 4). What is the effective collision diameter derived from these data using Eq. (7 - 46)?

Table 7 - 4 Kinetic Data for Nitrogen Decomposition					
Temperature, K	592	603.2	627	651.5	656
k(cc/mole-s)	522	755	1700	4020	5030

Used with permission from M. Bodenstein, *Z. Physik Chem.*, **100**, (1922), p. 106.

CHAPTER 8

SIMPLE AND COMPLEX RATE LAWS

FOR HOMOGENEOUS REACTIONS

The theories of homogeneous chemical kinetics quite naturally lead to the description of simple rate laws. For example, unimolecular reactions modeled by the Lindemann Theory may show rate laws of first to second order depending upon the total pressure of the system; whereas bimolecular reactions, as modeled by the collision theory, clearly show only second-order kinetics. In this chapter, phenomenological rate laws are presented and integrated to the appropriate form such that the model constants may be extracted from data. These homogeneous rate laws will provide the proper background to develop heterogeneous rate laws in Chapter 9.

Phenomenological Rate Laws of Irreversible Reactions

Among the simple reactions for study are those designated as "irreversible reactions." Since all reactions are reversible from the theory of microreversibility, irreversibility on the macroscopic scale denotes a class of reactions for which the reverse reaction rate is slow and the chemical equilibrium favors the products. As such, the reverse reaction may be neglected in modeling the actual process, thus simplifying the analysis.

The phenomenological rate law is a correlation of the observed reaction rates. Far from being a universal law, these equations are merely expressions of functional dependence of rate upon temperature and reactant/product partial pressures. In the most general sense, a rate law expresses the intensive reaction rate, r, as

$$r = r(c_i, T) \tag{8 - 1}$$

where

c_i = concentration of species i in partial pressures or in molarity, and so on

T = reaction temperature, usually in absolute temperature (but not always)

r = intensive reaction rate = R_i/ν_i
R_i = extensive reaction rate = $(1/V)d(N_i)/dt$
N_i = moles of species i
V = volume of reaction mixture
t = real time
ν_i = stoichiometric coefficient of species i

In some cases Eq. (8 - 1) may be simplified by assuming the temperature and concentration dependent functions are separable into the functions, k and g.

$$r = k(T)g(c_i) \tag{8 - 2}$$

Equation (8 - 2) shows the reaction order is defined by the function g; whereas the reactive cross sectional area, frequency factor, and activation energy terms are consolidated into the function, k. The traditional expression for k attributed to Arrhenius is

$$k = k_o \, exp(-E/RT) \tag{8 - 3}$$

where k_o is the preexponential factor and E is the activation energy in units consistent with the gas constant R. Although it is often taken as a constant independent of temperature, k_o is a function of temperature as described by several models in Chapter 7.

$$k_o = (8\pi RT/Nm)^{0.5}d^2, \text{ or} \tag{8 - 4}$$

$$k_o = (8\pi RT/Nm)^{0.5}d^2(E^{'}/RT)^{(n-1)}/(n-1)!, \text{ or} \tag{8 - 5}$$

$$k_o = w \, exp(-E^{'}/kT)/(n - 1)! \int u^{(n-1)} exp(-u) \, du/\{1 + w/z[M](u/[u + E^{'}/kT])^{(n-1)}\} \tag{8 - 6}$$

The expressions for k_o presented here range from the simple collision theory expression (8 - 4) to the complicated RRK expression (8 - 6). Over small temperature ranges (ca., 10-25°C), k_o is essentially constant next to the exponential term, for most systems.

We turn now to the models for the function, g. For a simple reaction between gaseous molecules involving only one elementary reaction step, the expression for g is a simple consequence of probability theory. The function, g, describes the probability of collision between the number of molecules (molecularity, m_i) required by the elementary reaction step. As the unimolecular reaction has been described in detail previously, we will focus on the bimolecular, gas phase reaction involving one elementary step between the reactant species A and B to form products:

$$A + B \rightarrow products$$

Here, the probability of collision, P, depends upon the product of the reactant concentrations each raised to the first power.

$$g = P = [A][B] \qquad (8-7)$$

This expression may be generalized as the following extended product over all reactant species showing stoichiometric coefficients, ν_i

$$g = \Pi[A_i]^{\nu_i} \qquad (8-8)$$

Equation (8 - 8) is generally known as the concentration dependent term of the "power law" kinetics for irreversible reactions. This equation combined with the rate constant term, k, yields the following expressions for zero-, first-, and second-order, irreversible kinetics of the reaction A + B → products.

zero order: $\qquad r = k \qquad (8-9)$

first order: $\qquad r = kc \quad$ *(one reactant)* $\qquad (8-10)$

second order: $\qquad r = kc^2 \quad$ *(one reactant)* $\qquad (8-11)$

$\qquad\qquad\qquad r = k[A][B] \quad$ *(two reactants)* $\qquad (8-12)$

shifting order: $\qquad r = kKc/(1 + Kc) \quad$ *(first to zero order)* $\qquad (8-13)$

The general expression for g, Eq. (8 - 8), may be extended to termolecular and higher-order reactions; however, these expressions will not be written here. These higher-order reactions (i.e., greater than three) in the gas phase are generally rare. The probabilities of gaseous, three-body collisions are small at normal gas densities; hence, the reaction rates are low.

Caution is advised in the use of Eq. (8 - 8) for all reactions: the assumptions inherent to its derivation may not apply to all reactions. The unimolecular decomposition is but one striking example of its misuse. The accepted reaction mechanism for thermally stimulated, unimolecular decompositions requires two elementary steps to describe the processes. The complete rate law for the unimolecular reactions is clearly not linear; however, the high pressure limit of this rate expression is the same as Eq. (8 - 10).

Integrated Rate Laws for Irreversible Reactions

One means for extracting kinetic constants from constant density batch reactor data is to integrate the rate law, linearize the resulting equation, and fit to data. Consider the general rate law given by Eq. (8 - 2),

$$r = (1/\nu_i)\, dc_i/dt = k(T)g(c_i) \tag{8 - 14}$$

The integrated rate law is

$$\int dc_i/[k(T)g(c_i)\nu_i] = \int dt \tag{8 - 15}$$

which demands a knowledge of the reaction temperature as a function of reactant concentration, c_i. By proper experimental design, one of the two extremes in batch reactor design (adiabatic or isothermal reactor) may be realized. For the case of isothermal reactor, Eq. (8 - 15) may be simplified to the following:

$$\int dc_i/[\nu_i g(c_i)] = k(T)t \tag{8 - 16}$$

Once the function, g, is known, the integration may be performed in principle. Consider the following example problem.

Example 1

Find the isothermal, integrated rate laws for zero-, first-, and second-order irreversible kinetics described by the equations (8 - 9, - 10, - 11, - 12) for the irreversible reaction:

$$aA \rightarrow products$$

Linearize these expressions and show how to extract rate constants, pre-exponential factors, and activation energies from data of reactant concentration versus time at two different temperatures.

Solution

First perform the integrations:

1. Zero order:

$$-dc = ak(T)\, dt$$

$$c = c_o - ak(T)t \tag{8 - 17}$$

2. First order:

$$- d\ln(c) = ak(T)\, dt$$

$$\ln(c_o/c) = ak(T)t \tag{8 - 18}$$

3. Second order, one reactant:

$$-dc/c^2 = ak(T)\ dt$$

$$1/c = 1/c_o + ak(T)t \qquad\qquad (8 \text{ - } 19)$$

4. Second order, two reactants:

$$aA + bB \rightarrow products$$

This solution is made easier by a change of variable from reactant concentrations to fractional conversion of the limiting reagent, f. Thus, for limiting reagent, A,

$$[A] = [A]_o - [A]_o f \qquad\qquad (8 \text{ - } 20)$$

$$[B] = [B]_o - (b/a)[A]_o f \qquad\qquad (8 \text{ - } 21)$$

and for the second-order rate law

$$r = - (1/a)\ d[A]/dt = k(T)[A][B]$$

Now making the substitutions called for in the rate law

$$(1/a)[A]_o\ df/dt = k(T)\{[A]_o(1 - f)\}\{[B]_o - [A]_o(b/a)f\}$$

which may be simplified to the following if $M = [B]_o/[A]_o$

$$df/dt = ak(T)[A]_o(1 - f)[M - (b/a)f]$$

Separation is now affected on this differential equation to give the following with $n = b/a$

$$df/\{(1 - f)[M - nf]\} = ak(T)[A]_o dt \qquad\qquad (8 \text{ - } 22)$$

A particularly useful procedure for integrating Eq. (8 - 22) is to expand the integrand of the dummy variable, f, as

$$df/\{(1 - f)[M - nf]\} = df\{ H/(1 - f) + J/[M - nf]\} \qquad\qquad (8 \text{ - } 23)$$

This partial fraction expansion has values of the constants, H and J, as follows

$$H = 1/(M - n) \quad and \quad J = 1/(1 - M/n) \qquad (8 - 24)$$

With this simplification, the integrated rate law is

$$-H \ln(1 - f) - (J/n) \ln[M - nf] = ak(T)[A]_o t \qquad (8 - 25)$$

Equations (8 - 17, - 18, - 19, - 25) are the linearized forms of the rate laws. The data must be processed as called for in each equation with the integrated ordinate plotted versus the time variable on the abscissa. Table 8 - 1 summarizes thèse results. The activation energy parameters (k_o and E_a) may be determined from these correlated data of $k(T)$ at two temperatures, T_1 and T_2, using the Arrhenius expression,

$$\ln[k(T)] = \ln[k_o] + (-E_a/RT) \qquad (8 - 26)$$

One may plot $\ln[k(T)]$ vs. $(1/T)$ for $T = T_1$, T_2 (using absolute temperature) to obtain a line. The intercept is $\ln[k_o]$, whereas the slope is $(-E_a/R)$. Using the value of R in the appropriate units, the activation energy is determined from the line slope. The units of k_o will be the same as those of k in the original data.

Table 8 - 1
Linearized Forms of Integrated Rate Equations, Example 1

Rate Law	Ordinate	Abscissa	Slope	Intercept
Zero	c	t	$ak(T)$	c_o
First	$\ln[c_o/c]$	t	$ak(T)$	0
Second, 1	$1/c$	t	$ak(T)$	$1/c_o$
Second, 2	$\ln[(M-nf)/(1-f)]$	t	$ak(T)[A]_o(M-n)$	0

Consider now the integrated expression for first law, irreversible kinetics. If a

$$\ln[c_o/c] = ak(T)t \qquad (8 - 27)$$

change of variable is made from concentration of reactant to fractional conversion of reactant, $f = \{c_o - c\}/c_o$, then the integrated rate law is

$$\ln[1/(1 - f)] = ak(T)t \qquad (8 - 28)$$

One may apply this expression to conversion versus time data for several different initial concentrations of reactant. Equation (8 - 28) shows the data for a first order reaction would describe only one line, independent of the initial concentration of the reactant. Any systematic departure from this one line as the initial concentration is changed indicates nonlinear kinetics. Consider the isotopic exchange of iodine between benzyl iodide and potassium iodide at room temperature.[1] The fractional exchange of iodine versus time in a batch reactor (Fig. 8 - 1) clearly shows the reaction is non-linear in reactant concentrations. That is, the rate of isotopic exchange changes with reactant concentrations. In the present case, increasing the reactant concentrations increases the rate of exchange with an overall order of reaction greater than unity. If the fractional exchange data were not affected by the reactant concentrations, then an overall order of first would be indicated. This technique for processing the raw data of conversion versus reaction severity constitutes the classical test for first-order, irreversible kinetics. As such, it is the most sensitive indicator for strictly linear kinetics.

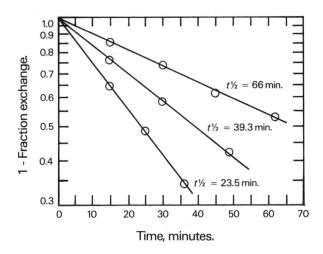

Figure 8 - 1 Isotopic Exchange of Benzyl Iodide with KI

Reprinted with permission from P. Stillson, and M. Kahn, *J. Am. Chem. Soc.* **75**, (1953), p. 3580. Copyright 1953, American Chemical Society.

Example 2

Find the integrated rate law for the shifting order rate law given by Eq. (8 - 13). Show how data may be used to determine the constants.

Solution

We illustrate a technique useful for integrating several types of shifting order rate laws. Equation (8 - 16) may be applied to the shifting order equations as follows:

$$-(1 + Kc)(dc)/(ac) = kK(dt) \qquad (8 - 29)$$

Notice two things about Eq. (8 - 29): the adsorption equilibrium coefficient, K, is a function of temperature and this function appears on the left-hand side of the equation mixed the concentration variable, c. If the kinetic data are isothermal, this equation can be formally integrated by dividing the left-hand side through as indicated and integrating term by term.

$$\int(1/a)(dc/c) + \int(K/a)\,dc = -kKt \qquad (8 - 30)$$

or upon integration we have

$$(1/a)\,\ln(c/c_o) + (K/a)(c - c_o) = -kKt \qquad (8 - 31)$$

where c_o is the concentration of the reactant at time $= 0$. Thus, the proper method of processing the data for this rate law is to plot the function $[\ln(c/c_o) + (K)(c - c_o)]$ versus t. The slope of this plot is $(-akK)$. Notice, the value of K must be known in order to construct the dimensionless plot. This suggests that the rate data must be manipulated first to extract the coefficient, K. Invert the rate expression Eq. (8 - 13) to yield

$$(1/r) = [1/(kKc)] + (1/k) \qquad (8 - 32)$$

Now multiply Eq. (8 - 32) by c to give one form of the linearized rate expression,

$$(c/r) = (1/kK) + (c/k) \qquad (8 - 33)$$

Data of (c/r) are plotted versus c to yield a slope equal to $(1/k)$ and the intercept is $(1/kK)$. Thus, eq. (8 - 33) may be used to find the parameters k and K; whereas the integrated rate Eq. (8 - 31) may be employed to check the value of k.

Ernst[2] describes a method to estimate the initial reaction rate, the order of reaction and the Langmuir coefficients without resorting to a trial and error procedure. This method was originally reported by Wilkinson[3] as a means to approximate any rate law, by a series, sum of power law models which have been truncated to retain the second-order terms of concentration and lower powers. The method is illustrated in the problems section at the end of the chapter.

Phenomenological Rate Laws of Reversible Reactions

Reversible reactions are those for which the observed reaction achieves equilibrium before the reactants are totally converted. The reaction rate laws must

Chapter 8 Homogeneous Rate Laws

therefore include the equilibrium condition of zero net rate at the equilibrium conversion, f_e. Thus, for one independent reaction the forward reaction rate, r_f, equals that of the reverse reaction, r_r, at the equilibrium condition

$$r_f(T, f_e) = r_r(T, f_e) \tag{8 - 34}$$

such that the net reaction rate is zero. The correlated, reversible rate expression is written for the disappearance of reactant, A, as

$$(-r) = r_f - r_r = k_f(T)g(f) - k_r(T)h(f) = k_f[g(f) - (1/K)h(f)] \tag{8 - 35}$$

where

$k_f(T), k_r(T)$ = forward and reverse rate constants = k, k'
$K(T)$ = $k_f(T)/k_r(T)$ = equilibrium constant
$g(f), h(f)$ = concentration dependent functions of rate expression

Now the task is to find the functional forms of $g(f)$ and $h(f)$ plus the constants, k_f and K, which best represent the data of the unknown system. Consider now the cases described for the irreversible kinetics for the reaction

$$A + B \rightarrow P$$

1. *zero order:* $r = k - k' = k(T)$ (8 - 36)

2. *first order:* $r = k\{[A] - (1/K)[P]\}$ (8 - 37)

3. *second order:* $r = k\{[A]^2 - (1/K)[P]^2\}$ *(one reactant)*   (8 - 38)

                           $r = k\{[A][B] - (1/K)[P]^2\}$ *(two reactants)*   (8 - 39)

Example 3

Find the integrated rate laws for zero-, first-, and second-order reversible kinetics described in Eqs. (8 - 36,- 37,- 38). Linearize these expressions and show how to extract rate constants, preexponential factors, and activation energies from data of reactant concentrations versus time at two temperatures.

Solution

1. zero order:

Since the rate expression for the reversible case has the same form as that for the irreversible case, the integrated rate law is the same as Eq. (8 - 17).

2. first order: $[P] = 0$ at $t = 0$:

$$[A]_o(df/dt) = [A]_o k\{(1 - f) - (1/K)(f)\} \qquad (8 - 40)$$

$$df/\{(1 - f) - (1/K)(f)\} = k \, dt \qquad (8 - 41)$$

$$ln\{1 - (1 + 1/K)f\} = -k(1 + 1/K)t \qquad (8 - 42)$$

Data of the equilibrium conversion, f_e, may be used to eliminate the equilibrium constant, K, from the integrated expression. At f_e, forward and reverse rates are equal; thus

$$(1 - f_e) = (1/K)f_e \qquad (8 - 43)$$

or

$$K = f_e/(1 - f_e) \qquad (8 - 44)$$

The integrated rate expression is

$$ln[1 - f/f_e] = -kt/f_e \qquad (8 - 45)$$

From Eq. (8 - 45) it is apparent that data of f versus t and the equilibrium conversion, f_e, are needed to correlate the equilibrium limited reaction rate data from a first-order, reversible reaction.

3. second order: $[P] = 0$ at $t = 0$; one reactant:

$$(df/dt) = k[A]_o\{(1 - f)^2 - (1/K)f^2\} \qquad (8 - 46)$$

This equation may be separated if the data are isothermal; that is, K is fixed for the one set of data.

$$df/\{1 - 2f + (1 - 1/K)f^2\} = k[A]_o \, dt \qquad (8 - 47)$$

This equation may be rearranged by completing the square in the denominator on the left-hand side of the equation to give

$$df/\{(1 - f)^2 - (f/K - 1)^2 - 2f/K + 1\} = k[A]_o t \qquad (8 - 48)$$

A useful technique for integrating the expression in f is to expand the left-hand side into three pieces using the partial fraction expansion procedure:

$$df/\{(1 - f)^2 - (f/K - 1)^2 - 2f/K + 1\} =$$
$$A\, df/(1 - f)^2 - B\, df/(f/K - 1)^2 + C\, df/(1-2f/K) \qquad (8 - 49)$$

The coefficients are evaluated in a straightforward manner to give

$$A = 1/\{1 - 2/K - ([1/K] - 1)^2\} \qquad (8 - 50)$$

$$B = -1/\{(1 - K)^2 - 1\} \qquad (8 - 51)$$

$$C = 1/\{(1 - K/2)^2 - (1/4)\} \qquad (8 - 52)$$

Now the term by term integration may be carried out to give the following:

$$A/(1 - f) + BK/[(f/K) - 1] - (CK/2)\ln\{1 - (2f/K)\} = k[A]_o t \qquad (8 - 53)$$

where f must be less than $K/2$ in the integration limits.

The rate constants are extracted from the data at finite reaction times; whereas the equilibrium constant is determined from the equilibrium conversion. The parameters are determined from each set of isothermal data. The rate constants are correlated versus temperature using the Arrhenius expression.

Isotopic Exchange Reactions

An important type of reversible reaction, the isotopic exchange reaction, has enjoyed great success in the elucidation of reaction mechanisms. Several attributes of the exchange reaction contribute to its popularity among kineticists.

1. Free energy of reaction is zero.
2. Enthalpy of reaction is zero.
3. No net driving force in gross chemical species so that the system is always at chemical equilibrium even though it may be far displaced from isotopic equilibrium.

One particularly useful isotope exchange reaction is between the two stable (nonradioactive) isotopes of hydrogen. Consider the H-D reactions of

$$RCH_3 + RCD_3 = RC(d_0, d_1, d_2, d_3) \qquad (8\text{-}54)$$

where R contains no exchangeable hydrogen species, and d_0 - d_3 denotes the number of deuteriums incorporated into the molecule. Although Eq. (8 - 54) specifically addresses gaseous intermolecular exchange, valuable mechanistic information may be extracted about surface catalyzed reactions.

Since the reactants/products of the isotope exchange reaction are chemically similar, special analytical techniques must be employed to follow the progress of the exchange. These tools include the mass spectrometer for the stable isotopes and counting instruments (Geiger-Muller counter, vibrating reed electrometer, or scintillation device) for the unstable isotopes. The isotopic labels afford a means to detect exchange processes which occur.

Exchange Kinetics

Consider the general isotope exchange reaction for which the quantity marked with the asterisk (*) is the labeled species.

$$AX + X^* = AX^* + X \qquad (8\text{-}55)$$

A mass balance may be performed to denote the concentration of species AX, labeled or unlabeled, as C_a and the concentration of species X, labeled or not, as C_x and the concentration of labeled species (AX^* and X^*) as C^*.

$$C_a = [AX] + [AX^*] \qquad (8\text{-}56)$$

$$C_x = [X] + [X^*] \qquad (8\text{-}57)$$

$$C^* = [AX^*] + [X^*] \qquad (8\text{-}58)$$

Using these balances we may estimate the statistical distribution of the labeled species between the two forms according to the concentrations. At equilibrium the concentration of species AX containing the label is just the fraction of total molecules that are AX and AX^* times the total concentration of labeled X^* molecules. In a similar manner, the concentration of X^* molecules at equilibrium is just the fraction of total molecules that are X^* and X times the total concentration of labeled X^* molecules.

$$[AX^*]_\infty = C^* \{C_a/(C_a + C_x)\} \qquad (8\text{-}59)$$

$$[X^*]_\infty = C^* \{C_x/(C_a + C_x)\} \qquad (8\text{-}60)$$

Several assumptions are appropriate to complete the derivation:
1. No isotope effect; therefore, $\Delta G = 0$ and $K = 1$.
2. No decay of X^* during the time of the experiment.
3. Only one exchange reaction is occurring.

A development now follows for which two molecular mechanisms are used in the description of the exchanges followed by a development which assumes no mechanism.

Mechanism 1

For this mechanism we write

$$AX = A + X \; ; AX^* = A + X^* \tag{8 - 61}$$

By assumption (1), these two dissociative reactions have identical rate constants. For a finite isotope effect, then K^* for the isotopic molecule would be different than that for the nonisotopic molecule. The rate of formation of AX^* is

$$d[AX^*]/dt = k_{-1}[A][X^*] - k_1[AX^*] \tag{8 - 62}$$

Now expressions for $[A]$ and $[X^*]$ are derived from C_a, C_x, and C^* as follows. The equilibrium relation for the decomposition of AX^* is

$$(k_1/k_{-1}) = [A][X^*]/[AX^*] \tag{8 - 63}$$

which may be rearranged to give the concentration of A explicitly,

$$[A] = (k_1/k_{-1})[AX^*]/[X^*] \tag{8 - 64}$$

At equilibrium the expressions for $[AX^*]$ and $[X^*]$ are substituted for in Eq. (8 - 62) to give

$$[A] = (k_1/k_{-1})C_a/C_x \tag{8 - 65}$$

The position of the equilibrium, heavily favoring the reactant AX^*, allows for the $[A]$ at any time t to be very similar to that at equilibrium. Thus,

$$d(AX^*)/dt = k_1 C^* C_a/\{C_a + C_x\} - k_1(AX^*) \tag{8 - 66}$$

Equation (8 - 66) may be integrated to give the following:

$$d(AX^*)/\{[AX^*]_\infty - [AX^*]\} = k_1 dt \tag{8 - 67}$$

$$ln\{[AX^*]_\infty - (AX^*)\} = -k_1 t + ln[AX^*]_\infty \tag{8 - 68}$$

Thus, we determine the forward rate constant k_1 from the equilibrium data of C_a and C_x and the time dependent data of $[AX^*]$. However, this derivation assumed a particular mechanism to calculate the rate constant. In the following two derivations the same form of the equation results, whatever the mechanism for the exchange.

Mechanism 2

Consider now a bimolecular exchange mechanism.

$$AX + X^* = AX^* + X \qquad (8 - 69)$$

for which the rate expression is

$$d[AX^*]/dt = k_2[AX][X^*] - k_2[AX^*][X] \qquad (8 - 70)$$

Now we appeal to Eqs. (8 - 56), (8 - 57), and (8 - 58) to give

$$C_a + C_x = [AX] + [AX^*] + [X] + [X^*] \qquad (8 - 71)$$

and

$$(C_a + C_x)[AX^*] = \{[AX] + [AX^*] + [X] + [X^*]\}[AX^*] \qquad (8 - 72)$$

and

$$(C_a + C_x)[AX^*]_\infty = C^* C_a = \{[AX] + [X^*]\}\{[AX] + [AX^*]\} \qquad (8 - 73)$$

If we subtract Eq. (8 - 72) from Eq. (8 - 73) the result is

$$(C_a + C_x)\{[AX^*]_\infty - [AX^*]\} = [AX][X^*] - [AX^*][X] \qquad (8 - 74)$$

Thus we may rewrite the apparent second-order rate law, Eq. (8 - 70), as a first-order equation

$$d[AX^*]/dt = k_2(C_a + C_x)\{[AX^*]_\infty - [AX^*]\} \qquad (8 - 75)$$

This may be integrated to give

$$ln\{[AX^*]_\infty - [AX^*]\} = -k_2(C_a + C_x)t + ln[AX^*]_\infty \qquad (8 - 76)$$

Notice Eqs. (8 - 68) and (8 - 76) are identical if k_1 is set equal to $k_2(C_a + C_x)$. The units of first- and second-order rate constants differ by that of concentration.

General Derivation

A general derivation of the exchange process may be developed beginning with a general expression for the exchange rate, R. This expression may involve events unproductive to changing the total concentration of tagged AX^* molecules for example, $AX + X$ and $AX^* + X^*$. Thus, we may write the effective exchange rate, R', as the fraction of the total rate.

$$R' = FR \qquad (8 - 77)$$

The task now is to derive the fraction F in terms of the concentrations of tagged and untagged species. This function accounts for only those exchanges which are effective in altering the concentration of AX^*. Let

$$[AX]/C_a = \textit{fraction of all AX species which are not labeled} \qquad (8 - 78)$$

and

$$[X^*]/C_x = \textit{fraction of all X species which are labeled} \qquad (8 - 79)$$

Thus, for the exchange between unlabeled AX and labeled X^* to produce AX^*, the fractional rate to produce AX^* is given by

$$(R'/R)_p = \{[AX]/C_a\}\{[X^*]/C_x\} \qquad (8 - 80)$$

whereas the fractional rate to destroy AX^* is given by

$$(R'/R)_d = \{[AX^*]/C_a\}\{[X]/C_x\} \qquad (8 - 81)$$

The net rate of formation of $[AX^*]$ is given by the difference of the rates of production and destruction.

$$R'/R = \{[AX][X^*] - [AX^*][X]\}/C_a C_x \qquad (8 - 82)$$

Notice the right-hand side of Eq. (8 - 82) is very similar to that of Eq. (8 - 74); therefore we have

$$d[AX^*]/dt = kR' = kR\{(C_a + C_x)/C_a C_x\}\{[AX^*]_\infty - [AX^*]\} \qquad (8 - 83)$$

This last equation may be integrated between zero time and t to give

$$ln\{1 - [AX^*]/[AX^*]_\infty\} = kR\{C_a + C_x\}/C_a C_x t \qquad (8 - 84)$$

This equation is first order in the approach of AX* to equilibrium concentration even though the overall exchange process has arbitrary functionality. The utility of this approach is to experimentally determine the functional dependence of R upon the reactant concentrations with the fewest data. It has been shown that the time required for which the argument of the log term is one-half may be conveniently related to R as follows:

$$R = C_a C_x / \{C_a + C_x\}\, ln(2)/t'$$ (8 - 85)

where t' is the time required to increase the $[AX\ ]$ to one-half its value at equilibrium.[4] This experiment may be repeated for several different values of C_a and C_x to produce the corresponding values of R. These R values are then correlated to produce the function R (see Problems 8 - 2, - 6).

Conclusions

Phenomenological rate laws were presented and integrated for the most commonly encountered cases. Techniques for linearizing and extracting the model constants were demonstrated. In many cases, these model constants will be compared to evaluate the reactivity of a particular system. For the cases of design work, these same rate laws may be used to represent the kinetics and thus must be integrated if the systems under consideration may be represented by a "one-dimensional" model. The kinetics of isotope exchange reactions were developed. These systems all show linear kinetics in the concentration of the isotopic species quite independent of the actual kinetics of the reactions. The immediate conclusion is that *all exchange reactions follow first-order kinetics!*

REFERENCES

1. Stillson, P. and M. Kahn, *J. Am. Chem. Soc.*, **75**, (1953), p. 3580.
2. Ernst, W. R. and M. Shoaei, personal communication.
3. Wilkinson, R. W., *Chemistry and Industry*, (1961), p. 1395.
4. McKay, H. A. C., *Nature*, **142**, (1938), p. 997; *J. Am. Chem. Soc.*, **65**, (1943), p. 702.

PROBLEMS

1. The isotopic exchange reaction between benzyl iodide and aqueous KI was studied using labeled KI. The activity of ^{131}I was determined in the organic and aqueous phases as a function of time. From the activity data, the fraction of iodine exchanged, F, was calculated (Table 8 - 2). Determine the exchange rate and the half-time (time required for $F = 1/2$).

Table 8 - 2
Kinetic Data for Isotopic
Exchange between Benzyl Iodide and KI, Problem 1

t, sec	1 - F
0	1.00
900	0.64
1500	0.48
2200	0.34

Temperature equals $27.3^\circ C$, concentrations of reactants (benzyl iodide and KI) equal 0.0102 and 0.00763 M, respectively. Reprinted with permission from P. Stillson and M. Kahn, *J. Am. Chem. Soc.*, **75**, (1953), p. 3580. Copyright 1953, American Chemical Society.

2. The bromination of acetone is believed to follow the mechanism

$$acetone \Leftrightarrow enol$$

$$enol + bromine \rightarrow bromoacetone + HBr \qquad (8 - 86)$$

For the conditions of excess acetone, the reaction is nearly zero order. The reaction progress may be determined spectrophotometrically since the bromoacetone shows very strong absorbance which is proportional to its concentration in solution (i.e., Beer's Law is observed). At $20.3^\circ C$ for 0.4 N sulfuric acid and with initial concentrations of acetone and bromine as 0.645 and 0.0193 mole/cubic dm, the following data were obtained (Table 8 - 3). Calculate the zero-order rate constant for these data.

Table 8 - 3 Absorbance of Bromoacetone vs Time

Time, ksec	0	0.6	1.2	3.6	4.8	6.0	9.0
Absorbance	0.201	0.257	0.313	0.558	0.665	0.683	0.683

Used with permission from James H. Espensen, *Chemical Kinetics and Reaction Mechanisms* (New York: McGraw-Hill Book Co., 1987), pp. 39-40.

3. The overall stoichiometry for the reaction of NO with hydrogen is

$$2 NO + 2 H_2 \rightarrow N_2 + 2 H_2O \qquad (8 - 87)$$

The data for this reaction were obtained at $1099^\circ K$ in a constant volume, batch reactor which was stirred well. Reactants were present initially in stoichiometric proportions. These kinetic data are reported as initial rates, expressed as change in total pressure with time, versus reactant partial pressure (Table 8 - 4).

    a. Determine the rate law.
    b. Calculate the rate constant and give its units.

Table 8 - 4
Reaction Rate Data for Reduction
of Nitric Oxide at 1099 K, Problem 3

P_H, Torr	P_{NO}, Torr	dP/dt, Torr/sec
289	400	0.160
205	400	0.110
147	400	0.079
400	359	0.150
400	300	0.103
400	152	0.025

Used with permission from I. Amdur and G. Hammes, *Chemical Kinetics: Principles and Selected Topics* (New York: McGraw-Hill Book Company. 1966), pp. 22-23.

    4. Ernst and Shoaei[2] extended a method originally developed by Wilkinson[3] to analyze the conversion, f, versus time, t, data from a batch reactor. In this method, power law kinetics of order n are used to fit the data and the fitting equation is truncated to the following form.

$$t/f = (1/K) + nt/2 \qquad\qquad (8 - 88)$$

where

    n = order of reaction
    K = intitial reaction rate.

Use Eq. (8 - 88) to find the initial reaction rate and order for the following data in Table 8 - 5.

    5. The shifting order kinetics modeled by Eqs. (8 - 13) and (8 - 31) may be approximated by the following equation.[2]

$$t/f = 1/R + (nt/2) \qquad\qquad (8 - 89)$$

Chapter 8 Homogeneous Rate Laws

Table 8 - 5 Conversion vs Time Data for Reaction of Unknown Order

Time, sec	0	20	40	60	80	100	200	300
Conversion	0.00	0.17	0.29	0.38	0.44	0.50	0.67	0.75

where

R = the initial rate at zero conversion

$n = 1/(1 + K'c_o)$

Fit the low conversion data in Table 8 - 6 to Eq. (8 - 89) to find $K'c_o$. Compare the fitted value to that used to generate the data: 0.5. Is this a viable and accurate method to determine the equilibrium adsorption coefficient?

Table 8 - 6 Conversion versus Time Data for Shifting Order Kinetics

Time, sec	0	76	155	238	323	507	711
Conversion	0.00	0.05	0.10	0.15	0.20	0.30	0.40

6. The exchange reaction described in Problem 1 was studied to describe the rate of exchange as a function of reactant concentrations. Table 8 - 7 shows the half-lives at $27.3°C$ for several reactant concentrations. Find the exchange rate for each data set and correlate the rate as a function of reactant concentrations.

Table 8 - 7 Reaction Data for Exchange of Iodine between Potassium and Benzene, Problem 6

Half-Life, sec	[BI], M	[KI], M
4350	0.00311	0.0025
2360	0.00672	0.00371
3960	0.00370	0.00219

Reprinted with permission from P. Stillson and M. Kahn, *J. Am. Chem. Soc.*, **75**, (1953), p. 3580 Copyright 1953, American Chemical Society.

CHAPTER 9

RATE LAWS FOR HETEROGENEOUS REACTIONS

The rate laws for heterogeneous reactions are written using the fundamentals developed for homogeneous reactions plus those developed for the adsorption/desorption events. A reaction mechanism is posed for the system, and the absolute rate theory is applied to each step of the mechanism. All steps, save for the rate limiting step(s), are written as equilibrium reactions and the resulting differential equation(s) is(are) solved. Examples are presented for the common mechanisms.

Langmuir-Hinshelwood (Hougen-Watson) Kinetics

The reactions at a gas/solid interface are modeled as a combination of two elementary reaction steps: an equilibrium between adsorbed reactant species and those in the gas phase followed by a kinetically controlled reaction on the surface involving adsorbed species. The sorbed products are assumed to be in equilibrium with the products in the gas phase. The rate laws for the reaction between the adsorbed species are assumed to be similar to that of the homogeneous analogs (e.g., power law kinetics of the reactants at the surface concentrations).

Consider now an energy diagram for the surface catalyzed reaction of A to form B (Fig. 9 - 1). First, gaseous A is sorbed onto the surface, liberating sorption energy, Q_a. The sorption process may experience an activation barrier of energy, E_1. The surface reaction surpasses the energy barrier, E_{tr}, to form chemisorbed product B and release the energy of reaction, ΔH. Chemisorbed B is released from the surface with the absorption of energy, Q_b. Thus, the observed energy barrier is

$$E_{obs} = E_{tr} - Q_a = E_{tr} + \Delta H_{ads} \qquad (9 - 1)$$

under the conditions that Q_a (a positive number) is significant. Equation (9 - 1) shows the observed activation energy can be different from the true activation energy if the heat of adsorption for the reactants is large.

The rate law for power law kinetics is an extended product over all species i:

$$r = k \prod_i (\theta_i)^{\nu_i} \qquad (9 - 2)$$

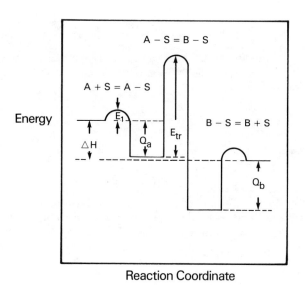

Reaction Coordinate

Figure 9 - 1   Energy Diagram for Surface-Catalyzed Reaction

where

$$\theta_i \quad = \text{fractional coverages of species } i$$
$$\nu_i \quad = \text{exponent in the rate law}$$
$$k \quad = \text{intrinsic rate constant}$$

Recall now from Chapter 3 the expressions for the fractional coverages in terms of reactant partial pressures in the gas phase.   Thus for one adsorbate, nondissociative sorption the rate law is

$$r = k\{Kp/(1 + Kp)\} \tag{9 - 3}$$

where

$$K \quad = \text{adsorption equilibrium constant, atm}^{-1}$$
$$p \quad = \text{partial pressure of reactant, atm}$$

Equation (9 - 3) is the most simple of the Langmuir-Hinshelwood equations. It clearly shows the "shifting order" behavior common to these equations. Consider the "low pressure" approximation to Eq. (9 - 3) where Kp is much less than 1, such that

$$r = kKp = k_{obs}p \tag{9 - 4}$$

Equation (9 - 4) shows first-order behavior and for the temperature response of k_{obs}, each term is differentiated as follows:

$$d(\ln k_{obs})/d(1/T) = -E_{obs}/R = d(\ln k)/d(1/T) + d(\ln K)/d(1/T) \qquad (9 - 5)$$

For each term in Eq. (9 - 5) we may substitute its equivalent assuming the rate constants behave according to the Arrhenius Rule and the adsorption equilibrium constants are described by Eq. (3 - 77).

$$- E_{obs}/R = -E_{tr}/R - \Delta H_{ads}/R$$

Thus solving for E_{obs} gives

$$E_{obs} = E_{tr} + \Delta H_{ads} = E_{tr} - Q_a \qquad (9 - 6)$$

which is the same as Eq. (9 - 1). Clearly, the observed values of the activation energy will differ from that of the intrinsic activation energy, E_{tr}, by Q_a. A better estimate of the true activation energy is available using the rates determined at high coverages. Here the rate is given by

$$r = k$$

since $Kp \gg 1$. At this condition the observed rate constant is the true rate constant and true activation energy. For all intermediate pressures (hence coverages), then

$$E_{tr} > E_{obs} > E_{tr} - Q_a \qquad (9 - 7)$$

This development may be extended to other sorption stoichiometries. Consider a reaction mechanism where two reactants (A and B) are adsorbed associatively and then react by a bimolecular reaction. Suppose that the product (C) is held only weakly on the surface. Table 3 - 1 suggests the proper expression of θ_i for this competitive adsorption among A, B, and C is

$$\theta_i = K_i p_i/(1 + K_a p_a + K_b p_b + K_c p_c) \qquad (9 - 8)$$

However, if we assume component C is held in a weak fashion on the surface then $K_c p_c$ is insignificant next to all of the other denominator terms. The irreversible reaction rate expression for power law kinetics, Eq. (9 - 2), is as follows

$$r_i = k K_a p_a K_b p_b/(1 + K_a p_a + K_b p_b)^2 \qquad (9 - 9)$$

assuming the rate is first order in each of the reactant fractional coverages.

Consider now the behavior of Eq. (9 - 9) for classical reaction order experiments where isothermal reaction rates are measured as one partial pressure is varied and all others are held constant. For varying partial pressure of reactant A,

Chapter 9 Heterogeneous Rate Laws

$$r = (kK_bp_bK_a)p_a/[(1 + K_bp_b) + K_ap_a]^2 \qquad \text{(9 - 10)}$$

Equation (9 - 10) has been grouped such that the constant-value terms are enclosed in parentheses and the terms varying with p_a are set apart. The response of the rate with increasing p_a and constant p_b shows a maximum in the rate at the value of p_a given by

$$p_a\big|_{max} = (1 + K_bp_b)/K_a \qquad \text{(9 - 11)}$$

Equation (9 - 11) results from an appeal to the calculus at the maximum value of r for which

$$(dr/dp_a)\big|_{max} = 0$$

Since Eq. (9 - 9) is symmetric with respect to A and B then the same form of the solution exists for reactant B with the obvious permutation of subscripts:

$$p_b\big|_{max} = (1 + K_ap_a)/K_b \qquad \text{(9 - 12)}$$

The ratio of $(p_a/p_b)\big|_{max}$ suggests a relationship between the partial pressures at the maximum as follows:

$$p_a/p_b\big|_{max} = \{1 + K_bp_b\}K_b/\{(1 + K_ap_a)K_a\} \qquad \text{(9 - 13)}$$

Parameter Estimation

The parameters for these equations (k, K_a, and K_b) are amenable to curve-fitting of the linearized equations. For the present case, return to Eq. (9 - 10) and invert each side to yield

$$(1/r) = [(1 + K_bp_b) + K_ap_a]^2/[(kK_bp_bK_a)p_a] \qquad \text{(9 - 14)}$$

If we extract the square root of Eq. (9 - 14) to give

$$(1/r)^{0.5} = [(1 + K_bp_b) + K_ap_a]/[(kK_bp_bK_a)p_a]^{0.5} \qquad \text{(9 - 15)}$$

The linearized form is obtained by multiplying by $p_a^{0.5}$ and rearranging the right-hand side of Eq. (9 - 15).

$$(p_a/r)^{0.5} = (1 + K_bp_b)/(kK_bp_bK_a)^{0.5} + [(K_a/kK_bp_b)^{0.5}]p_a \qquad \text{(9 - 16)}$$

Thus, a plot of $(p_a/r)^{0.5}$ versus p_a gives a straight line of slope $(K_a/kK_bp_b)^{0.5}$ and the intercept is $(1 + K_bp_b)/(kK_bp_bK_a)^{0.5}$. In practice, these data would be collected at low conversions to minimize the effects of product inhibition and for several different initial partial pressures of A at constant temperature and partial pressure of B. This would be repeated at another temperature thus allowing a description of the temperature response of K_a and k. The same set of experiments may be repeated to extract K_b at each temperature by varying the partial pressure of B only.

It must be remembered this development assumed the Langmuir-Hinshelwood equations accurately depicted the reaction mechanism. For an actual study, the validity of the mechanism must be established first. White and Hightower[1] showed the kinetics of propylene oxidative dimerization could be *adequately* explained by two mechanisms: the Langmuir-Hinshelwood mechanism and that given by Mars and van Krevelen.[2] As indicated in Reference 1, the Langmuir-Hinshelwood (L-H) model is not all inclusive of reaction rate models and is certainly *not* exclusive either.

Rideal-Eley Model

A simple modification of the L-H model allows for a description of a different class of reactions at interfaces. Here one reactant attacks the chemisorbed species without itself becoming chemisorbed. This mechanism of direct gas phase attack on a chemisorbed reactant was advanced by Rideal and Eley.[3] The rate law for reactants A and B is given by

$$r = k\theta_a(p_b)^n \qquad (9 \cdot 17)$$

Thus, a test for this mechanism is to hold reactant A concentration constant for an isothermal series of increasing partial pressures of reactant B. Usually, the logarithmic treatment of data (plot $\ln[r]$ versus $\ln[p_b]$) allows for the determination of the functional dependence of the rate upon concentration of B. The functional dependence of the rate upon the concentration of reactant A is handled just as demonstrated for the Langmuir-Hinshelwood kinetics.

Specialized Mechanisms - Mars-van Krevelen[2]

We have discussed general types of mechanisms which were formulated for the arbitrary reaction over the general catalyst. The results obtained from fitting data to these types of models range from good to disappointing. Models specifically detailed for the system at hand often show good agreement between predictions and the data. The Mars-van Krevelen model is frequently used to represent hydrocarbon oxidation reactions on oxide catalysts. The mechanism for this model assumes:

1. The catalyst is reduced by the reactant(s) and is subsequently reoxidized by molecular oxygen.
2. Both the reactants and the oxygen are adsorbed onto the surface.
3. The adsorption of reactants and products is in equilibrium to the gas phase reactants/products.
4. The rate of catalyst reduction is equal to that of catalyst reoxidation.

Thus, we may define the following:

n_r = rate of reduction = $k_1(p_h{}^n)\theta$

n_o = rate of reoxidation = $k_2(p_o{}^m)(1 - \theta)$

p_h = partial pressure of reactant

p_o = partial pressure of oxygen

θ = fractional coverage of reactant

and the stoichiometry of the oxidation of the reactant requires that

$$n_r = (1/b)n_o \qquad (9 - 18)$$

At steady state

$$bk_1p_h{}^n\theta = k_2p_o{}^m(1 - \theta) \qquad (9 - 19)$$

for which the fractional coverage may be extracted. Using this value for θ and substituting into the expression for n_r we have for the linearized form

$$1/n_r = [1/(k_1p_h{}^n)] + [b/(k_2p_o{}^m)] \qquad (9 - 20)$$

Equation (9 - 20) suggests a method for testing the validity of the Mars-van Krevelen model. The reciprocal rate of reactant disappearance is plotted versus reciprocal oxygen partial pressure to the mth power. Additional data to determine hydrocarbon order may be tested against reciprocal rates versus reciprocal hydrocarbon partial pressure to the nth power. Other researchers show fractional powers of both reactant and oxygen partial pressures may be necessary to fit the data.[1]

Specialized Mechanisms - Temkin-Pyzhev[4]

Several facts peculiar to the ammonia synthesis reaction led to the development of Temkin-Pyzhev kinetics. It was known that the rate of NH_3 formation was the same as the rate of N_2 adsorption and that the deuterium exchange reactions with NH_3 and H_2 over the ammonia catalysts were at equilibrium. These findings led to a reaction mechanism having the following assumptions:

1. The dissociation of N_2 was the kinetically slow step.
2. Nitrogen atoms are the predominant species on the surface.
3. Hydrogen and ammonia do not influence the rate of nitrogen adsorption.
4. Activation energies E_a, E_d for adsorption, desorption decrease linearly with surface coverage, θ.
5. The Elovich equation describes the rates of adsorption, desorption.

Thus, the rates of adsorption, desorption are written as follows:

$$r_a = k_ap(N_2) \exp(-g\theta) \qquad (9 - 21)$$

$$r_d = k_d \exp(+h\theta) \tag{9 - 22}$$

where

k_a, k_d = frequency factors for adsorption, desorption
p = nitrogen partial pressure
g, h = Elovich constants

The net rate of nitrogen uptake is equal to the rate of ammonia synthesis, r; thus

$$r = r_a - r_d = k_a p \exp(-g\theta) - k_d \exp(+h\theta) \tag{9 - 23}$$

Consider now the N_2 pressure (called the *virtual* nitrogen pressure) required by the formula. Temkin *defines* this virtual pressure as that for the equilibrium

$$N_2\big|_{ads} + 3H_2 \Leftrightarrow 2NH_3$$

where

$$K = p(NH_3)^2/[p(N_2)p(H_2)^3]$$

or

$$p(N_2) = [1/K][p(NH_3)^2/p(H_2)^3]$$

Next, expressions for θ are determined by equating the rates of adsorption and desorption as

$$(k_a/K)[p(NH_3)^2/p(H_2)^3]\exp(-g\theta) = k_d \exp(h\theta)$$

or

$$\theta = [1/(g + h)]ln\{(k_a/k_d)[p(NH_3)^2/(p(H_2)^3 K)]\} \tag{9 - 24}$$

The synthesis rate is given by

$$r = [p(NH_3)]^2 k_a/[K[p(H_2)]^3]\{[(K_a/k_d p(NH_3)]^2/[K[p(H_2)]^3]\}^{-(g/a)}$$
$$- k_d\{(k_a/k_d)[p(NH_3)]^2/[K[p(H_2)]^2]\}^{(h/a)} \tag{9 - 25}$$

where

$$a = g + h$$

Several examples of heterogeneous reactions will help clarify these points.

Example 1   **Kinetics of Heterogeneous Reactions; OXDD of Propylene over Bi₂O₃**

Reaction of propylene in O_2 over Bi_2O_3 was studied in a microcatalytic flow reactor between $748°$ and $898°K$. Low conversion data at different inlet concentrations of reactants were used to define initial reaction rates at each temperature (Fig. 9 - 2). The initial conversion rates were plotted versus reciprocal temperature to define apparent activation energies of 22.0 kcal/mole (Fig. 9 - 3). Initial rates of O_2 consumption at $823°K$ were plotted versus O_2 concentration at one propylene concentration to define the O_2 order (Fig. 9 - 4). The same procedure was repeated to define the propylene order (Fig. 9 - 5). All of these data were fit to two models: the Langmuir-Hinshelwood and the Mars-van Krevelen models.

The Langmuir-Hinshelwood Model

$$-r\big|_{oxygen} = k_2\theta_p^2\theta_o = k_2(K_pC_p)^2(K_oC_o)/[1 + K_pC_p + K_oC_o]^3$$

where

$$K_p = 37.8 \text{ M}^{-1} \exp\{-(1250/R)[823 - T]/823T\}$$
$$K_o = 34.0 \text{ M}^{-1} \exp\{-(2500/R)[823 - T]/823T\}$$

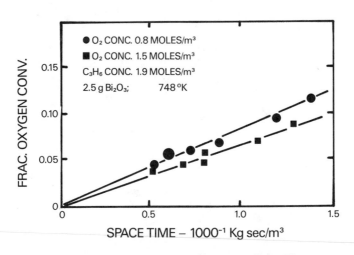

Figure 9 - 2   Oxygen Conversion versus Space-Time

Used with permission from M. G. White and J. W. Hightower, *J. Catal.* **82**, (1983), p. 185.

Chapter 9 *Heterogeneous Rate Laws*

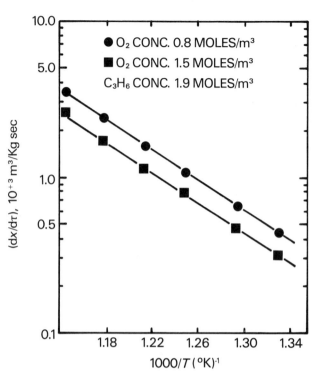

Figure 9 - 3 Arrhenius Plot of OXDD of Propylene over Bismuth Oxide

Used with permission from M. G. White and J. W. Hightower, *J. Catal.* **82**, (1983), p. 185.

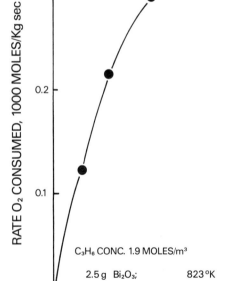

Figure 9 - 4 Initial Rates of Oxygen Consumption versus Oxygen Concentration

Used with permission from M. G. White and J. W. Hightower, *J. Catal.* **82**, (1983), p. 185.

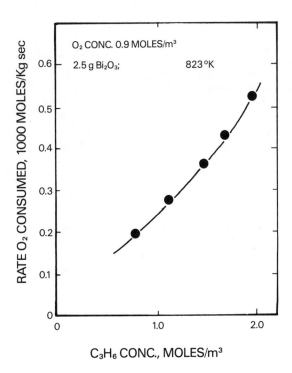

O₂ CONC. 0.9 MOLES/m³

2.5 g Bi₂O₃;   823 °K

C₃H₆ CONC., MOLES/m³

Figure 9 - 5 Initial Rates of Oxygen Consumption versus Propylene Concentration

Used with permission from M. G. White and J. W. Hightower, *J. Catal.* **82**, (1983), p. 185.

The rate constant k_2 is determined from a fit of the integrated rate equation versus $(K_pC_p)^2K_ot$ as in Fig. 9 - 6. Here the constants for the ordinate are as follows:

$$m = (1 + K_pC_p')$$
$$A = K_oC_o'$$
$$x = \text{fractional } O_2 \text{ conversion}$$
$$t = \text{space-time in the reactor}$$

and the superscript (') indicates inlet concentration of reactant. These fitted values of k_2 are represented on an Arrhenius plot, Figure 9 - 7, to determine the best value of E_a (21.4 kcal/mole).

The Mars-van Krevelen Model

$$-(1/r\,|_{oxygen}) = 1/[k_1(C_p)^{0.5}] + b/[k_2(C_o)^{1.5}]$$

This equation was fit to the same reaction rate data at each temperature to yield the Arrhenius plot for k_2 shown in Figure 9 - 8. The value for the activation energy, E_a, was 21 kcal/mole.

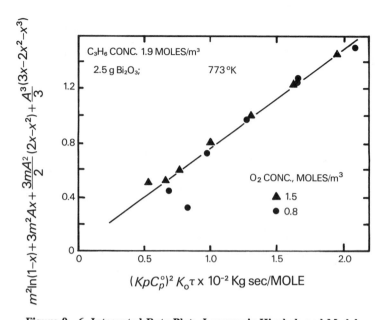

Figure 9 - 6 Integrated Rate Plot: Lamgmuir-Hinshelwood Model

Used with permission from M. G. White and J. W. Hightower, *J. Catal.* **82**, (1983), p. 185.

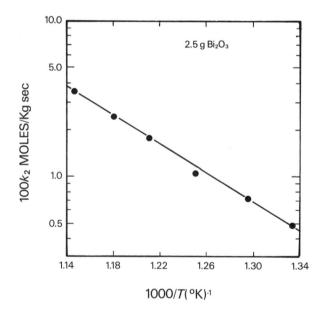

Figure 9 - 7 Arrhenius Plot for Integrated Langmuir-Hinshelwood Model

Used with permission from M. G. White and J. W. Hightower, *J. Catal.* **82**, 185 (1983).

Chapter 9 Heterogeneous Rate Laws

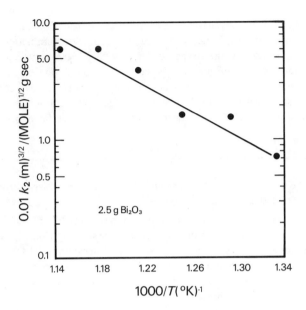

Figure 9 - 8 Arrhenius Plot for Integrated Mars-van Krevelen Model

Used with permission from M. G. White and J. W. Hightower, *J. Catal.* **82**, (1983), p. 185.

Even though it correlates the data very well, the Langmuir-Hinshelwood model has at least two serious shortcomings which contradict the current theories in oxide catalysis. First, it assumes associative oxygen adsorption, and second, the adsorption equilibrium constants increase slightly with temperature instead of decreasing as theory predicts they should.

For correlation purposes, this form of the Langmuir-Hinshelwood equation predicts OXDD reaction rates with oxygen orders varying between -2 to +1 as the oxygen partial pressure is decreased. Thus, it can be used to explain the results reported by Seiyama et al.[5] (negative order in oxygen), Swift et al.[6] (zero order in oxygen), Trimm et al.[7, 8] (positive fractional order in oxygen) and Pashegorova et al.[9] (first order in oxygen).

We explain our rate expression by recognizing the fate of molecular oxygen. Oxygen is consumed to form selective products, nonselective products, and for catalyst reoxidation. The site regeneration may be either a dissociative process[8] or an associative process[10] followed by disproportionation of S-O_2 and S-(vacant) to form two S-O species. The path to total combustion products may involve either (or both) chemisorbed or lattice oxygen.[11] Since our microreactor results showed that 1,5-hexadiene could be oxidized to CO_2 in the absence of gaseous oxygen, lattice oxygen must be involved. The path to selective products is thought to involve lattice oxygen, which is a process generally modeled by zero-order oxygen kinetics, although some researchers[12] have postulated a hydrocarbon peroxide ion mechanism for the selective oxidation of propylene, a related reaction. The summation of all these contributions indicates that the oxygen order may vary between zero and first order, depending upon which rate process is controlling. Our rate data, Figure 9 - 4, support this conclusion, although it is not possible to assign relative contributions to each of these selective, nonselective, and site regeneration processes. In short, our rate equation does not rule out any mechanisms that have been proposed for the reaction.

Example 2   **Kinetics of Heterogeneous Reactions; Selective Propylene Oxidation over Cu-Sn Oxides[13]**

Product poisoning can influence reaction rates. Useful information regarding surface states may be extracted from such poisoning studies. Consider the inhibition of acrolein production rates by products water and acrolein (C_3H_4O) over a copper oxide catalyst and one modified by SnO (Fig. 9 - 9). The incomplete poisoning of either catalyst by water shows at least two different sites produce acrolein. Moreover, the character of the sites is different.

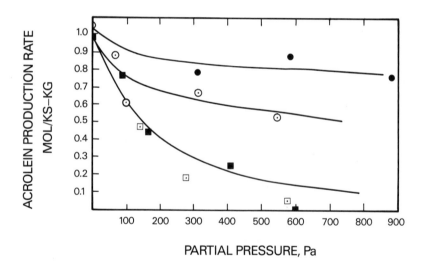

PARTIAL PRESSURE, Pa

Figure 9 - 9 Inhibition of Reaction Rate by Products

Inhibition of the reaction rate by products over a copper-tin oxide (1/1) catalysts and a copper oxide catalyst. List of symbols: water inhibition data over copper oxide, ●; water inhibition data over copper-tin oxide, ○; acrolein inhibition data over copper oxide, ■; acrolein data over copper-tin oxide, □. Solid lines are predicted reaction rate using Eq. (9 - 26). Reaction temperature is 548 K. Used with permission from D. E. Self, J. D. Oakes, and M. G. White, *Am. Inst. Chem. Eng. J.*, **29**, No. 4, (1983), p. 625.

The data of product inhibition by water and acrolein over Cu_2O and copper-tin oxide catalysts show that both catalysts are poisoned "equally" by acrolein, but the copper oxide catalyst is less sensitive to water poisoning than the mixed oxide catalyst (Fig. 9 - 9). The rate data for acrolein inhibition over each catalysts are correlated for acrolein partial pressures less than 300 Pa and water pressures to 900 Pa, by the following equation:

$$r_a = kP_o/[\{1 + K_aP_a + K_wP_w^{0.5}\}]\{1/(1 + K_aP_a)\} \qquad (9 - 26)$$

At $598°K$ the correlated constants are as follows for the copper oxide and the copper-tin oxide catalysts (Table 9 - 1).

At acrolein partial pressures greater than 300 Pa this correlation will not predict the slightly "negative" acrolein production rate observed for the reaction over cuprous oxide. The equilibrium adsorption coefficient for the water in contact with the Cu_2O catalyst is roughly 12.6% of the water adsorption coefficient for the copper-tin oxide catalysts. The correlated water equilibrium coefficients show the copper-tin oxide catalysts are very sensitive to water sorption.

Table 9 - 1 Correlated Rate and Equilibrium Constants

Constants	Copper Oxide	Copper-Tin Oxide
kP_o [mol acrolein/ks-kg]	1.01	1.01
K_w [Pa^{-1}]	1.55×10^{-4}	1.23×10^{-3}
K_a [Pa^{-1}]	2.48×10^{-3}	2.48×10^{-3}

Used with permission from D. E. Self, J. D. Oakes, and M. G. White, *Am. Inst. Chem. Eng. J.*, **29**, No. 4, (1983), p. 625.

The difference in reaction orders between the water inhibition rate data and the acrolein inhibition rate data over each of the catalysts may be explained by a two-site model in which acrolein and carbon dioxide are produced on each site. One site is more selective to the acrolein reaction than the other site. A dual-site model for copper oxide catalysts was reported by Mikhal'chenko et al.[14] These Russian investigators showed that the active copper surface sorbs propylene and acrolein in a weak, reversible manner by π electron bonding on one type of site and to sorb on another type of site in a strong, "irreversible" manner. Water was reported to be strongly sorbed on the surface. In the Russian model, however, the nonselective reactions occur on one site (the strong, irreversible sorbing sites) and the selective reactions occur on the weakly sorbing, reversible sites. The type A site of our model is poisoned by sorption of polar molecules such as water; whereas the type B site is poisoned by molecules having π electrons. From a consideration of the correlated rates and adsorption equilibrium constants for the acrolein inhibition data over the copper-tin oxide and Cu_2O catalysts, both have nearly the same total number of acrolein producing sites per unit mass of catalysts, assuming the acrolein molecule is sorbed to both type A and B sites. The water inhibition data suggest the copper oxide catalyst have only 12% of the type A sites as the copper-tin oxide catalysts. We speculate the type B sites are less selective than the type A sites since the catalyst having the greater number of A sites, the mixed oxide, is the more selective catalyst.

With these few data it would be premature to characterize these sites completely. We may speculate the type B site sorbs acrolein and propylene via the π electron system, whereas the type A site may sorb species by some other mechanism in addition to π bonding, possibly through ionosorption as described by Hauffe.[15] Such sorption by electron withdrawing groups (H_2O, OH^-, O^-, and so on) is favored on Cu_2O and p-type semiconductors in preference to electron donating groups.[16] Water, having no π electrons, will not sorb to the type B site; thus, the rate of acrolein production will not be completely inhibited by the sorption of water. On the other hand, acrolein may sorb to both types of sites to completely inhibit the production of acrolein. On the A sites, the very polar acrolein molecule (dipole moment = 2.90 DeBye at $298^{\circ}K^{17}$) may sorb by ionosorption; whereas on the type B site it will sorb via the conjugated π electron system.

Compensation Effect

In heterogeneous catalysis the compensation effect is illustrated by changes in the Arrhenius constants such that the observed rate constant changes but little[18] for a series of related reactions over a catalyst or for one reaction over a series of catalysts. Consider the case of a reaction catalyzed by a series of similar catalysts. The response of the correlated rate constants with temperature for each catalyst, given by Figure 9 - 10, shows a point of intersection at temperature, T'. For these data, it may said that compensation exists. This behavior, observed for a number of systems, has been represented by the following relation:

$$lnA = aE + b = aE + lnk' \tag{9 - 27}$$

where

A = preexponential factor
E = activation energy
a, b = compensation constants

Equation (9 - 28) may be rearranged for the *ith* catalyst as

$$lnA_i = aE_i + lnk_i \tag{9 - 28}$$

which may be expressed as

$$A_i = k' \, exp[aE_i] = k' \, exp[E_i/RT] \tag{9 - 29}$$

Since all the curves in Figure 9 - 10 intersect at temperature, T', then the Arrhenius expressions for all the catalysts may be related as

$$k_i = A_i \, exp[-E_i/RT'] = k_j = A_j \, exp[-E_j/RT']$$ (9 - 30)

Constants A_i, A_j may be eliminated using Eq. (9 - 29) to give

$$k' \, exp[-E_i(1/RT' - a)] = k' \, exp[-E_j(1/RT' - a)]$$ (9 - 31)

Clearly, the activation energies are not equal; thus, the condition imposed by Eq. (9 - 31) must be kept only when

$$a = 1/RT'$$ (9 - 32)

The new form of the Arrhenius relation is

$$k_i = k' \, exp[-E_i(1/T - 1/T')/R]$$ (9 - 33)

The phenomenon of compensation may be explained if a heterogeneous site distribution is assumed. For one catalyst we write

$$k = A \sum n_i \, exp[-E_i/RT]$$ (9 - 34)

where according to the site distribution, n_i sites have energy, E_i. Now if the relationship between n_i and E_i is postulated as

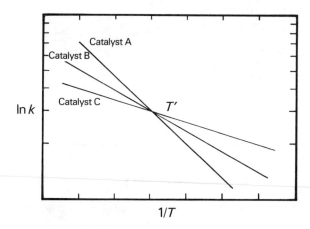

Figure 9 - 10 Compensation Plot

$$n_j = a \, exp[-E_j/b]$$ (9 - 35)

then the new expression for the rate constant is

$$k = A \, \Sigma a \, exp[-E_j/b] \, exp[-E_j/RT]$$ (9 - 36)

For closely spaced energy levels, E_j, the summation in Eq. (9 - 36) may be replaced by an integral to give

$$k = Aa \int exp[-E(1/RT - 1/b)] \, dE = Aa/(1/RT - 1/b) \, exp[-E \, (1/RT - 1/b)]$$ (9 - 37)

Examination of Eq. (9 - 37) shows it is similar to Eq. (9 - 33) if b equals with RT' and if k' equals $Aa/(1/RT - 1/RT')$.

Example 3   Compensation in Mixed, Metal Oxide Catalysts

We have reported compensation effects in the selective propylene oxidation to acrolein over promoted copper oxide catalysts.[13]   Figure 9 - 11 illustrates the

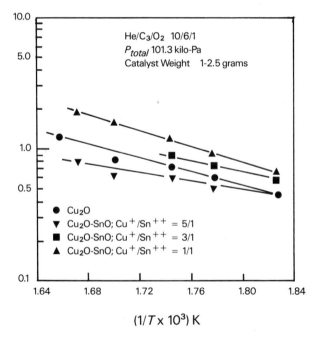

$$(1/T \times 10^3) \, K$$

Figure 9 - 11 Compensation in Rate Constants for Selective Oxidation of Propylene

Used with permission from D. E. Self, J. D. Oakes, and M. G. White, *Am. Inst. Chem. Eng. J.,* **29**, No. 4, (1982), p. 625.

Chapter 9 Heterogeneous Rate Laws

compensation effect for several promoted catalysts having tin incorporations of 0 to 50 atom% and Table 9 - 2 summarizes the Arrhenius constants. The primed (') constants are specific to the weight of copper in the sample; whereas the unprimed constants are specific to the total catalyst weight.

The energetics of the selective oxidation reaction appear to vary with catalyst composition in such a way as to suggest the reaction is compensated over these catalysts. There is good agreement between the observed activation energy (55 kJ/mol) over the Cu_2O and the data of Margolis[19] for Cu_2O/silica and Cu_2O/pumice (E_a = 50-59 kJ/mol). A sensitivity analysis of our data shows an uncertainty of ± 10.8 kJ/mol assuming a relative error of 10%. This uncertainty in the activation energy does not falsify the apparent compensation effect.

Table 9 - 2 Arrhenius Constants for Mixed Oxide Selective Oxidation Catalysts

Catalyst	Atom% Tin	Wt.% Tin	E_a [kJ/mol]	k_o	k_o'
Cu_2O	0	0	54.8	4.3×10^{28}	4.3×10^{28}
Cu_2O/SnO (5/1)	16	25	29.7	1.9×10^{26}	2.5×10^{26}
Cu_2O/SnO (3/1)	33	25	47.2	1.1×10^{28}	1.8×10^{28}
Cu_2O/SnO (1/1)	50	66	60.6	7.4×10^{29}	2.5×10^{29}

Used with permission from D. E. Self, J. D. Oakes, and M. G. White, *Am. Inst. Chem. Eng. J.*, **29**, No. 4, (1982), p. 625.

The specific surface areas of the catalyst are nearly equal and unchanged by the standard pretreatment; thus, no error is incurred for comparing the reactions rates on a per unit mass basis rather than a per unit surface area basis. The reaction is not hindered by pore diffusion, the Arrhenius plots are linear, and the Weisz-Prater criterion is satisfied.[20]

The preexponential factors specific to the weight of copper ion (Table 9 - 2) show a decrease as the copper ion is diluted in the inert SnO from 16 to 50 atom% tin. However, only the most dilute catalyst shows a preexponential factor greater than pure Cu_2O. It is possible that a small number of copper ions are "dissolved" into the SnO lattice during the activation of the catalysts. A compensation plot of our data over four catalyst compositions (Fig. 9 - 12) is in good agreement to the correlated data of Moro-Oka[21] for propylene oxidation over several mixed oxides. The compensation parameters of this work (a = .1017, b = 15.7) are nearly identical to the data of Moro-Oka (a = .1017, b = 15.5). One interpretation of these data is a surface heterogeneous to sorption and/or reaction. In light of the previous discussion of the dual-site model, the compensation due to product sorption is a reasonable explanation of the rate constant data.

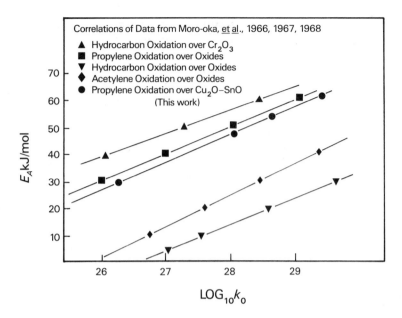

Figure 9 - 12 Compensation in the Rate Constants for Oxidation Reactions

Used with permission from D. E. Self, J. D. Oakes, and M. G. White, *Am. Inst. Chem. Eng. J.,* **29**, No. 4, (1983), p. 625.

Conclusions

This brief review of several selected heterogeneous reaction rate laws illustrates several points:

1. The rate laws and the data to support them are never sufficient to prove any mechanism suggested by the rate law. Almost every study of kinetics must be augmented by data from other sources (such as isotopic tracers, and so on). These other studies are designed to support assumptions inherent to the rate law.
2. Often two rate laws, based on entirely different mechanisms, may explain the kinetic data equally well. In this case, it is encumbent upon the investigator to speculate which mechanism best explains all the data known about the system or related systems.

It is rare for any rate law/mechanism to explain all the data perfectly. The scientist, often, must compromise the agreement between all the data and the chosen mechanism for the sake of time and money.

REFERENCES

1. White, Mark G. and Joe W. Hightower, *J. Catal.*, **82**, (1983), p. 185.
2. Mars, P. and D. W. van Krevelen, *Spec. Suppl. Chem. Eng. Sci.*, **3**, (1954), p. 41.
3. Thomas, J. M. and W. J. Thomas, *Introduction to the Principles of Heterogeneous Catalysis* (New York: Academic Press, Inc., 1981).
4. Temkin, M. I. and V. Pyzhev, *Acta Physicochim. USSR*, **12**, (1940), p. 327.
5. Seiyama, T., T. Uda, I. Mochida, and M. Egashira, *J. Catal.*, **34**, (1974), p. 29.
6. Swift, H., J. E. Bozik, and J. A. Ondrey, *J. Catal.*, **21**, (1971), p. 212.
7. Trimm, D. L., and L. A. Doerr, *J. Catal.*, **23**, (1971), p. 49.
8. Trimm, D. L., and L. A. Doerr, *J. Catal.*, **26**, (1972), p. 1.
9. Pashegorova, V. S., L. M. Kaliberdo, and G. G. Lebedeva, *Neftekhimiya*, **16**, (1976), p. 840.
10. Mamedov, E. A., E. G. Gamaid-zade, F. Agaev, and R. G. Rizaev, *Kinet. Katal.*, **20**, (1979), p. 410.
11. Gamid-zade, E. G., E. A. Mamedov, and R. G. Rizaev, *Kinet. Katal.*, **20**, (1979), p. 405.
12. Keulks, G. W., L. D. Krenzke, and T. M. Notermann, *Adv. Catal.*, **27**, (1978), p. 183.
13. Self, D. E., J. D. Oakes, and M. G. White, *Am. Inst. Chem. Eng. J.*, **29**, No. 4, (1983), p. 625
14. Mikhal'chenko, V. G., V. D. Sokolovskii, and G. K. Boreskov, *Kinet. Katal.*, **14:3**, (1973), p. 602; **14:5**, (1973), p. 1099.
15. Hauffe, K., *Adv. Catal.*, **7**, (1955), p. 213.
16. White, M. G. and J. W. Hightower, *Am. Inst. Chem. Eng. J.*, **27**, (1981), p. 545.
17. Dean, J. A., Ed., *Lange's Handbook of Chemistry*, 12th ed. (New York: McGraw-Hill Book Co., 1979).
18. Bond, G. C., *Catalysis by Metals* (London: Academic Press, , 1962), p. 140.
19. Margolis, L. Ya., *Adv. Catal.*, **14**, (1963), p. 429.; Margolis, L. Ya., *Catal. Rev.*, **8:2**, (1973), p. 241.
20. Weisz, P. B., and C. D. Prater, *Adv. Catal.*, **6**,
21. Moro-Oka, Y. and A. Ozaki, *J. Catal.*, **5**, (1966), p. 116; *J. Catal.*, **10**, (1968), p. 84.
22. Beckler, R. K. and M. G. White, *J. Catal.*, **110**, (1988), p. 364.

PROBLEMS

1. The initial rate data for benzene hydrogenation over a Ni/kieselguhr catalyst are given in Table 9 - 3. For these experiments, excess hydrogen was used and the data are reported as the hydrogen partial pressure divided by the initial rate of

Table 9 - 3 Reaction Rate Data for Benzene Hydrogenation over Ni/Kieselguhr, Problem 1

Y_B	Temperature, $^\circ$C	$P_H/r_B \times 10^7$
0.02	86	20
0.04	86	20
0.06	86	20
0.08	86	20
0.10	86	20
0.02	124	5
0.04	124	4
0.06	124	4
0.08	124	4
0.10	124	4
0.01	155	2
0.02	155	1.67
0.04	155	1.33
0.06	155	1.25
0.08	155	1.17
0.10	155	1.14
0.01	173	1.54
0.02	173	1.11
0.03	173	0.927
0.04	173	0.833
0.06	173	0.745
0.08	173	0.71
0.10	173	0.667

benzene consumption for several temperatures and benzene mole fractions in the feed (Y_B). Determine the Langmuir-type rate expression which fits the data best.

a. Extract rate constants, equilibrium adsorption constants as functions of temperature.

b. Determine the activation energy from the rates constants, the preexponential factor and the enthalpy of adsorption plus the entropy of adsorption for benzene.

2. Verify the integrated rate ordinate in Figure 9 - 6, Example 1 by integrating the Lanmuir-Hinshelwood rate equation to give the following expression:

Chapter 9 Heterogeneous Rate Laws

$$m^2 \ln(1-x) + 3m^2Ax + (3/2)mA^2(2x - x^2) + (1/3)A^3(3x - 2x^2 - x^3) = k(K_P C_P{}^o)^2 K_o t \quad (9 - 38)$$

3. Beckler[22] reports the cyclopropane isomerization over supported polynuclear metal complexes is modeled by the following rate expression.

$$-r = kK[CP]/(1 + K[CP] + K'[P]) \quad (9 - 39)$$

where

$[CP], [P]$ = molar concentrations of cyclopropane and propylene, respectively

K, K' = equilibrium adsorption constants for cyclopropane and propylene, respectively

k = intrinsic rate constant for surface reaction

Integrate this expression and fit it to the following data of the reaction over a supported iron complex to determine k and K'. In a separate experiment, Beckler showed that the fractional conversion versus time did not change as the initial concentration of cyclopropane was increased. What can you conclude from these data about the relative magnitudes of K and K'?

Table 9 - 4
Conversion Data for Isomerization of
Cyclopropane over a Supported Iron Complex, Problem 3

Time, ksec	0	0.6	2.4	7.2	10.8	14.4	18.0	21.6	28.8	39.6	50.4	57.6
Conversion	0	0.01	0.04	0.12	0.17	0.22	0.27	0.31	0.39	0.49	0.57	0.61

4. Using the Wilkinson method modified by Ernst, calculate the initial reaction rate the data in Problem 3 and compare it to the following data for an equal weight of the supported chromium catalysts.

Table 9 - 5
Conversion Data for Isomerization of
Cyclopropane over a Supported Chromium Complex, Problem 4

Time, ksec	0	0.6	2.4	7.2	10.8	14.4	18.0	21.6	28.8	39.6	50.4	57.6
Conversion	0	0.01	0.03	0.10	0.15	0.19	0.23	0.27	0.35	0.44	0.51	0.55

5. Integrate the rate equation for the selective propylene oxidation to acrolein (Eq. 9-26) and process the following batch reactor data of oxygen conversion versus time. Extract the rate constant at the temperature of the data.

Table 9 - 6

Reaction Data for Selective Oxidation of Propylene
to Acrolein over a Copper-Tin Catalyst, Problem 5

Time, min	0	10	15	20	25	30	35	40	45	50	55	60
P, acrolein	0	0.16	0.23	0.36	0.47	0.57	0.70	0.73	0.80	0.87	0.92	1.00

$P_{ox}^{\circ} = 18.5\ Torr;\ P_{c3}^{\circ} = 52.5\ Torr;\ T = 150^{\circ}C$

Used with permission from J. D. Oakes, M. S. thesis, Georgia Institute of Technology (1983).

CHAPTER 10

ISOTOPICALLY LABELED COMPOUND STUDIES

Kinetic studies are necessary but never sufficient to establish a reaction mechanism. Most often, the kinetics suggest several mechanisms for which other techniques may be used to discriminate among them. We showed how the electrical properties[1] were useful to confirm the adsorption of O_2 in the OXDD of propylene over Bi_2O_3 at reaction conditions. Studies using isotopically labeled reactant compounds are useful to confirm/deny assumptions of a proposed reaction mechanism. In this section we discuss the use of ^2H, ^{13}C, ^{14}C, and ^{18}O labeled compounds. Each of the labeled compounds shows characteristics peculiar to that group.

Some general comments are appropriate concerning the tactics of using isotopically labeled compounds. The isotopic compound may used in "static" or "kinetic" experiments to elucidate a reaction mechanism. In the static experiment, the extent of isotope incorporation in the products and other reactants allows certain conclusions to be drawn regarding the reaction mechanism. An example will help to clarify the meaning of a "static" experiment. Consider the reaction mechanism for the hydrogenation of ethylene over a catalyst. An appropriate isotopic compound, deuterium, may be used to characterize the mechanism of the hydrogenation as molecular addition or as activation of the hydrogen molecule followed by stepwise addition of the hydrogen atoms to the ethylene. The products of a molecular addition of deuterium would be ethane-d_2 to the exclusion of other deuterated ethane species. The products of the stepwise addition would be a distribution of deuterated species: ethane-d_i, $i = 0$ to 6. These product slates would be fulfilled if the following assumptions were true:

1. Hydrogen and deuterium do not exchange with each other in the gas phase, homogeneously.

2. Reactant/product hydrocarbons do not exchange hydrogens for deuteriums, homogeneously.

These assumptions are but further clarification/specification of the reaction mechanism which may be confirmed by additional experiments. The homogeneous exchange of hydrogens for deuteriums may be studied by a separate reaction of equimolar mixture of the hydrogen- and deuterium-containing reactants in the absence of a catalyst at the temperature and pressure of interest. If no exchange product is detected in the mass spectrometer (e.g., HD if the reactants are hydrogen and deuterium) then, homogeneous reactant scrambling is absent from the system of interest. In the present case three separate tests for H/D scrambling would be necessary:

1. Scrambling of H/D in an equimolar mixture of hydrogen/deuterium.
2. Scrambling of H/D in an equimolar mixture of the reactant ethylene-d_0 and ethylene-d_4 (perdeuterated ethylene).
3. Scrambling of H/D in an equimolar mixture of product ethane-d_0 and ethane-d_6 (perdeuterated ethane).

If no homogeneous exchange occurred after an overnight, noncatalytic test, then scrambling of H/D will not confuse the results of the test. These tests are necessary to confirm the integrity of the reactants prior to the reactive event and that of the product after the reactive event. Should these tests show no scrambling, the deuterium incorporation of the products from the studies over the catalyst will be characteristic of the catalytic reaction only.

Having eliminated homogeneous scrambling of H/D, one may check for catalytic scrambling by repeating these experiments with a catalyst present. If stepwise addition is to occur via activation of the hydrogen molecule, this activation step should be apparent even in the absence of the ethylene co-reactant. If the activiation of hydrogen is not apparent by the first test with a catalyst, then the activation must occur by a Rideal-Eley mechanism of gas phase hydrogen/deuterium reacting with a chemisorbed ethylene fragment to produce an HD and a deuterated hydrocarbon. For stepwise addition, the reactants and products may encounter the catalyst surface many times before leaving the reactor. If an atomic deuterium or hydrogen is present to react and/or exchange, then a statistical incorporation of deuterium is expected in the product ethane. This distribution may be predicted from combinatorial theory and the initial concentration of deuterium in the reacting gas (see Problem 10 - 1).

The "kinetic" experiment differs from the "static" experiment in that the rate of disappearance of the deuterium-containing reactant is compared to that of the hydrogen-containing reactant. For the following discussion, the species are hydrocarbons. In the kinetic experiments, one is trying to associate the rate limiting step with carbon-hydrogen bond scission. The kinetic differences in the rates are most obvious between a C-H and a C-D bond because of the relative differences in the reduced masses of the two systems: approximately 1 versus 2. One could study kinetic isotope effects for bonds involving other atoms; however, the magnitude of the effect would be much smaller than that observed between carbon and hydrogen because the relative, reduced masses would be very close to each other. Consider ^{12}CO and ^{13}CO. The reduced masses are 7.2 and 7.54, respectively. For this reason, one most often studies kinetic isotope effects for C-H bond scission reactions only. The genesis of the isotope effect arises from the effect of the reduced mass on the strength of the critical bond for oscillation along the critical reaction coordinate. The next section provides a brief review of the pertinent literature.

Deuterium Labeled Compounds

One of the most powerful tools for characterizing the rate determining step in reaction mechanisms of hydrocarbons is the deuterium labeled compound. We shall concentrate on the use of labeled hydrocarbons in characterizing carbon-hydrogen bond scissions. Consider a mechanism for which a carbon-hydrogen bond activation is *assumed* to be the kinetically slow step. This assumption may be confirmed/denied by a simple experiment using a mixture of deuterium-labeled hydrocarbon and the hydrogen form and searching for an isotope effect. Before describing the experiment, a discussion of the isotope effect is necessary.

Primary Kinetic Isotope Effect[2a, 2b]

Consider the reaction

$$A + B \Leftrightarrow M' \rightarrow products$$

proceeding through the transition state to the products. From transition state theory, the equilibrium constant describing the concentration of the activated complex, $[M']$, is

$$K = ([M']/[A][B])(\alpha'/\alpha_A \alpha_B) \qquad (10 - 1)$$

where α is an activity coefficient, $[A]$ and $[B]$ are reactant concentrations. The reaction rate, r, is

$$r = v'[M'] = v'K[A][B](\alpha_A \alpha_B/\alpha') \qquad (10 - 2)$$

where v' is the frequency for decomposition of the activated complex. The rate constant, k, may be written as

$$k = v'K(\alpha_A \alpha_B/\alpha') \qquad (10 - 3)$$

Recall from transition state theory (Chapter 7) that we may express K in terms of the partition functions (per unit volume) of all the molecular species.

$$K = (q'/V)/[(q_A/V)(q_B/V)] \, exp(-E_o/RT) \qquad (10 - 4)$$

and for v'

$$v' = (RT/Nh)q/q' \qquad (10 - 5)$$

Thus, the rate constant expression is

$$k = v'K = (RT/Nh)(q/V)/[(q_A/V)(q_B/V)] \exp(-E_o/RT) \tag{10-6}$$

for $\alpha' = \alpha_A = \alpha_B = 1$. The partition function for all modes other than vibration for the complex is q. Consider the rate constant, k^*, for an isotopically labeled A molecule,

$$k^* = (RT/Nh)(q^{*}/V)/[(q^{*}_A/V)(q^{*}_B/V)] \exp(-E_o/RT) \tag{10-7}$$

If we ratio Eq. (10 - 6) to Eq. (10 - 7) we have

$$k/k^* = (q'/q^{*})(q^{*}_A/q_A)(q^{*}_B/q_B) \exp[(E_o^* - E_o)/RT] \tag{10-8}$$

For most isotopically labeled reactants $E_o^* = E_o$; thus,

$$(k/k^*) = (q'/q^{*})(q^{*}_A/q_A)(q^{*}_B/q_B) \tag{10-9}$$

As usual the evalution of the partition functions proceeds by assuming the functions are separable. For molecule A we may write the translation, rotation, vibration, and electronic terms as given in Table 10 - 1.

Table 10 - 1
Partition Functions

Mode	Expression
Translation	$(2\pi m_i RT/N_o)^{1.5}/h^3$
Rotation	
Linear	$8\pi^2 IRT/N_o^2 sh$
Non linear	$8\pi^2(8\pi^3 I_x I_y I_z)^{0.5}(RT/N_o)^{1.5}/sh$
Vibration, nonlinear	$\exp[N_o h\nu/2RT](1 - \exp[-h\nu N_o/RT])$

I = moment of inertia = $1/[2\mu d^2]$
I_x, I_y, I_z = moments of inertia
s = rotational symmetry number
μ = reduced mass = $m_A m_B/(m_A + m_B)$
d = distance between the masses m_A & m_B
R = gas constant in energy units
T = absolute temperature
N_o = Avogadro's number
h = Planck's constant
ν = vibrational frequency

Thus, we may use these terms to express k/k^* as follows for a non linear molecule.

$$\{(s/s^*)(s_A^*/s_A)\}^{0.5}k/k^* = (\mu/\mu^*)^{1.5}(I_xI_yI_z)/[(I_xI_yI_z)/(I_x^*I_y^*I_z^*)]^{0.5} \quad x$$

$$\prod_{i=1}^{3N'-7}\{1 - exp[-N_ohv_i^*/RT]\}/\{1 - exp[-N_ohv_i/RT] \quad x$$

$$\prod_{i=1}^{3N-6}\{1 - exp[-N_ohv_{iA}/RT]\}/\{1 - exp[-N_ov_{iA}^*/RT]\} \quad x$$

$$\prod_{i=1}^{3N'-7}\{exp[(N_oh/2RT)(v_i - v_i^*)]\}\prod_{i=1}^{3N-6}\{exp[(N_oh/2RT)(v_{iA}^* - v_{iA})]\} \qquad (10 - 10)$$

Here N' and N are the vibrational modes of the complex and the A molecule. Recall one mode of vibration for the complex, the decomposition mode, has been included; thus the number of modes is $3N' - 7$. It is possible to apply Eq. (10 - 10) to the problem of deuterium-labeled hydrocarbons. For these compounds, the mass increase for a molecule with i deuteriums is given by

$$m^* = m + i \qquad (10 - 11)$$

where m is the mass of the molecule in proton form and m^* is the mass of the deuterated form. Moreover, this mass increase represents a small fraction of the original mass (less than 10%) even for a perdeuterated sample such that the symmetry numbers, and moments of inertia for the deuterated, undeuterated samples are very similar. Thus, Eq. (10 - 10) simplifies to

$$k/k^* = \{(m_A + i + m_B)(m_Am_B)(m_A + i)/(m_A + m_B)(m_A + i)(m_Am_B)\}^{1.5} \quad x$$

$$\prod_{i=1}^{3N'-7}\{1 - exp[-N_ohv_i^*/RT]\}/\{1 - exp[-N_ohv_i/RT]\} \quad x$$

$$\prod_{i=1}^{3N-6}\{1 - exp[-N_ohv_{iA}/RT]\}/\{1 - exp[-N_ohv_{iA}^*/RT]\} \quad x$$

$$\prod_{i=1}^{3N-6} exp[\mu/2(v_i^* - v_i)]/\prod_{i=1}^{3N-6} exp[\mu/2(v_{iA}^* - v_{iA})] \qquad (10 - 12)$$

where

$$\mu = N_oh/RT$$

$$k/k^* = \{mass\ effect\} \times \{excitation\ terms\} \times \{zero\text{-}point\ energy\} \qquad (10 - 13)$$

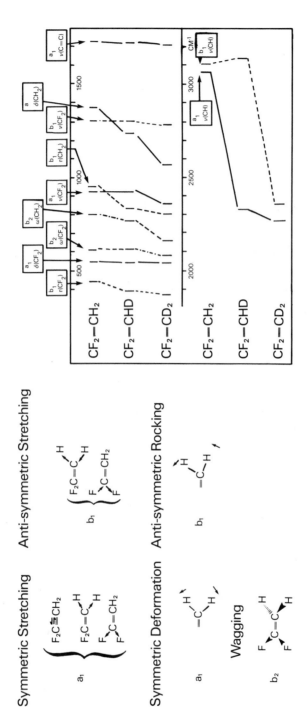

Figure 10 - 1   Vibrational States for Difluoroethylene

Used with permission from W. F. Edgell and C. J. Ultee, *J. Chem. Phys.*, **22**, (1954), p. 1983.

where

mass effect =
$$\{(m_A + m_B + i)/(m_A + m_B)\}^{1.5}$$

excitation terms =
$$\{1 - exp[-\mu\nu_i^*]\}/\{1 - exp[-\mu\nu_i]\}\{1 - exp[-\mu\nu_{iA}]\}/\{1 - exp[-\mu\nu_{iA}^*]\}$$

zero-point energy =
$$exp[\mu/2(\nu_i^* - \nu_i)]/exp[\mu/2(\nu_{ia}^* - \nu_{ia})]$$

Usually the mass effect is insignificant next to that of the zero-point energy and excitation terms. Considering the large values of $\mu\nu_i$ for most hydrocarbons, we see that $exp(-\mu\nu_i)$ is small. Thus the excitation terms make no significant contribution. Finally, we examine the zero-point energy terms. If we rewrite the zero-point energy (ZPE) terms as

$$ZPE = exp\{-\mu/2[(\nu_i - \nu_i^*) - (\nu_{ia} - \nu_{ia}^*)]\} \qquad (10 - 14)$$

As an example of the magnitude of the zero-point energy, consider the IR spectrum[2b] of ethylene difluoride with the hydrogens substituted by deuteriums: $CF_2=CH_2$, $CF_2=CHD$, and $CF_2=CD_2$. These molecules show their IR active vibrations designated a_1, b_1, and b_2. Consider the perdeuterated species; the a_1 vibrations are the symmetrical C-D, C-F, and C=C stretching modes plus the CF_2 and CD_2 deformation modes. The b_1 vibrations are the antisymmetric C-D and C-F stretching and the rocking of the CF_2 and CD_2 groups. The b_2 vibrations are the in- and out-of-plane wagging of the CF_2 and CD_2 groups.

Figure 10 - 1 shows these vibrations for the three compounds and Table 10 - 2 summarizes the frequencies of the C-D and C-H vibrations. Thus, if $\nu_i - \nu_i^*$ are the same in the transition state as in the reactant molecule for a_1 deformation, b_1, and b_2 relaxations, the only term that survives is the a_1 stretch. Using the frequencies of this

---

Table 10 - 2
Vibrational Frequencies of Difluoroethylene

Vibration	ν_i	ν_i^*	$(\nu_i - \nu_i^*)$ cm⁻¹
Stretch, a_1	3060	2260	800
Deformation, a_1	1380	1090	290
Stretch, b_1	3100	2350	750
Rock, b_1	950	820	130
Wag, b_2	800	650	150

Used with permission from W. F. Edgell and C. J. Ultee, *J. Chem. Phys.*, **22**, (1954), p. 1983.

stretch for the deuterated and nondeuterated species one may calculate the isotope effect using Eq. (10 - 14). The difference in frequencies between the deuterated and nondeuterated species is 3060 - 2260 = 800 cm^{-1}. Thus, the isotope effect is as follows when the temperature is 303°C:

$$k'/k = exp[(-h/2kT)(800cm^{-1})(3 \times 10^{10}\ cm/sec)] = exp[-1.93] = 0.145 \qquad (10 - 15)$$

The test for a primary isotope effect proceeds through a series of experiments usually carried at nominal reaction temperatures in a batch recirculation reactor or a microcatalytic flow reactor. The first test may involve equimolar mixtures of perdeuterated and undeuterated hydrocarbons to describe the isotope effect. If a primary isotope effect exists, the relative rates of reactant consumption (d_i to d_o) will be less than unity. The perdeuterated isotope is preferred so that all possible points of attack may be measured.

Consider the classical work of Adams and Jennings for selective propylene oxidation at 450°C over bismuth molybdate to test the allylic carbon hydride abstraction mechanism.[3a, 3b] The proposed mechanism to form the allyl radical was represented by the following elementary equation where S is a surface site:

$$H_2C = CH-CH_3 + S \rightarrow H_2C-CH-CH_2 + SH \qquad (10 - 16)$$

If Eq. (10 - 16) is the rate determining step, then a primary isotope effect should be observed for propylene labeled on the methyl end. The authors proposed a series of tests using equimolar mixtures of the undeuterated propylene and either 3-propylene-d_1, 1-propylene-d_1, or perdeuterated propylene as the co-reactants. For perdeuterated propylene the ratio of the rate constants, k'/k, was 0.55 ± 0.05, indicating a significant primary isotope effect. With this information it was possible to calculate k'/k for the tests using the monodeuterated propylene as the co-reactant. For example, the probability of abstracting a deuterium from the methyl is 1/3; thus the ratio of rate constants is

$$k'/k = (k^* + 2k)/3k = [(k^*/k) + 2]/3 \qquad (10 - 17)$$

Using $(k^*/k) = 0.55$, they calculated for 3-propylene-d_1 an isotope effect of

$$k'/k = (0.55 + 2)/3 = 0.85 \qquad (10 - 18)$$

If the allyl mechanism is operative then there should be no isotope effect when 1-propylene-d_1 is the co-reactant. Table 10 - 3 summarizes the results of Adams and Jennings for propylene oxidation over bismuth molybdate. Based on these results the authors conclude the kinetically slow step is allylic hydride abstraction. Subsequent study described the location of the deuterium atoms in the acrolein product[3, 4, 5] to confirm the allylic mechanism.

Table 10 - 3
Isotopic Rate Constants for Propylene Oxidation over Bismuth Molybdate at 450°C

Compound	Observed	Calculated [†]
C_3H_6	1	1
$CH_2 = CHCH_2D$	0.85 ± 0.02	0.85
$CHD = CHCH_3$	0.98 ± 0.02	1
C_3D_6	0.55 ± 0.05	-

[†] Ratios calculated using experimental results of perdeuterated test. Used with permission from C. R. Adams and T. J. Jennnings, *J. Catal.*, **3**, (1964), p. 546.

Unfortunately, the results of deuterated compounds studies are not always so convincing. For the OXDD of propylene to 1-5 hexadiene and benzene[6] a large primary isotope effect was observed at 600°C to confirm the rate limiting step as C-H cleavage (Fig. 10 - 2). The isotope effect is the ratio of the limiting slopes of propylene conversion with space-time for the C_3'-d_0 isotope to the C_3'-d_5 isotope. Moreover, only a small amount of H/D scrambling among reactant and product molecules occurs. The skewed distribution of d_i in products Figure 10 - 3 indicates there must be nonequilibrated isotopic scrambling along the path to form benzene. We summarized the mechanism as shown in Figure 10 - 4 where the darkness of the arrow showed the relative reaction rates to the respective products.

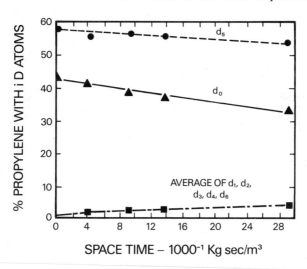

Figure 10 - 2 Hydrogen-Deuterium Scrambling/Isotope Effects: C_3H_6/C_3HD_5 Mixture at 873°K

Used with permission from M. G. White and J. W. Hightower, *J. Catal.*, **82**, (1983), p. 185.

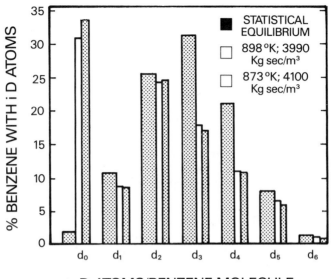

Figure 10 - 3   Hydrogen-Deuterium Distribution in Benzene

Used with permission from M. G. White and J. W. Hightower, *J. Catal.,* **82**, (1983), p. 185.

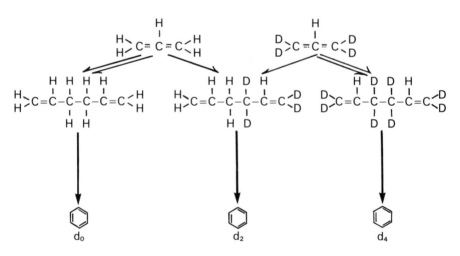

Figure 10 - 4   Reaction Mechanism for OXDD Reaction

Used with permission from M. G. White and J. W. Hightower, *J. Catal.,* **82**, (1983), p. 185.

Chapter 10 Isotopic Tracers

Labeled Carbon Compounds

The stable (^{13}C) and radioactive (^{14}C) isotopes of hydrocarbons have been used to study a number of reactions including the selective oxidation of propylene over bismuth molybdate. Consider the reaction of propylene labeled by ^{14}C to form acrolein.[7]

$$C_3H_6 + 1/2\ O_2 \rightarrow C_3H_4O + H_2O \qquad (10\text{ - }19)$$

Three possible mechanisms have been advanced to describe this reaction over bismuth molybdate. The initial step in each mechanism is as follows:

[A-1] $H_3C^*\text{-}CH_2 = CH_2 + O \rightarrow O = C^*H\text{-}CH = CH_2 + 2H;$ (I)

[A-2] $H_3C^*\text{-}CH_2 = CH_2 + O \rightarrow C^*H_2 = CH\text{-}C = O + 2H;$ (II)

[B] $propylene \rightarrow CH_2\text{-}CH\text{-}CH_2 + O \rightarrow$ equal amounts of I and II

Here, the asterisk denotes this carbon as the radioactive species so that the mechanisms may be distinguished easily one from another. By mechanisms A-1 and A-2, propylene reacts with a surface oxygen atom in a concerted manner to lose two hydride atoms and gain an oxygen atom. If the methyl atom loses the hydrides and the oxygen attaches there, species I will be formed by A-1. On the other hand, if the methyl and methylene carbons lose a hydride by some unspecifed mechanism, and if the methylene carbon receives the oxygen, then species II is formed by the A-2 mechanism. Mechanism B assumes a stepwise loss of hydrides to form the symmetric allyl radical followed by loss of the second H, and addition of the oxygen atom. The allyl radical is formed by loss of the hydrogen from the methyl carbon; the intermediates lead to equal amounts of I and II.

These reaction mechanisms may be used to predict the labeled products of reacting radioactive propylene over the catalyst. 1-Propylene-1-C* leads to the formation of species II, exclusively by mechanism A-1; species I, exclusively by mechanism A-2; and equal amounts of species I and II, by mechanism B. 1-Propylene-3-C* leads to the formation of species I, exclusively by mechanism A-1; species II, exclusively by mechanism A-2; and equal amounts of I; and II by mechanism B. 1-Propylene-2-C* forms species III, $CH_2 = C^*H\text{-}C = O$ by all mechanisms.

The radioactivity of CO and ethylene was analyzed after the product acrolein was photochemically decomposed:

$$CH_2 = CH\text{-}CHO + h\nu \rightarrow CO + CH_2 = CH_2 \qquad (10\text{ - }20)$$

In this manner the *position* of the radioactive carbon in the acrolein may be deduced from the ^{14}C in the CO and ethylene. The acrolein decomposition products, CO and ethylene, are easily separated by gas chromatography to be analyzed for specific

Table 10 - 4
Carbon-14 Labeled Propylene Studies of the Selective Oxidation over Bismuth Molybdate at 450°C

Labeled Compound	Specific Activities			Sum	Pred. Values of $R^{\dagger}$			Expt'l, R
	(C_3')	(CO)	(C2')		(A-1)	(A-2)	(B)	
1-propene-1-C*	0.99	0.45	0.49	0.94	∞	0.	1.	0.91 ± 0.13
1-propene-1-C*	0.99	0.44	0.48	0.92	∞	0.	1.	0.91 ± 0.20
1-propene-1-C*	0.40	0.17	0.19	0.35	∞	0.	1.	0.91 ± 0.01
1-propene-2-C*	0.25	0.01	0.24	0.25	0.	0.	0.	0.03 ± 0.01
1-propene-3-C*	0.50	0.25	0.25	0.50	0.	∞	1.	0.95 ± 0.12

† R is the specific activity of CO/specific activity of ethylene. Used with permission from W. M. H. Sachtler and N. H. DeBoer, *3rd Inter. Cong. Catal.* (Amsterdam: North-Holland Publ., Inc., 1964), Vol. I, p. 252.

radioactivities by counting devices. The ratio of the specific activities (R) carbon monoxide to ethylene is reported in Table 10 - 4 for predicted and experimental runs. Decomposition of the species I gives R equal to infinity, whereas species II leads to R equal to zero. The experimental results of the material balances, column 1 versus the sum of 2 and 3 in Table 10 - 4, show satisfactory closures of the balances for all the experiments. For each run, the experimental values of R are explained best by mechanism B. Values of R were predicted assuming no oligomer or metathesis products are formed and assuming the photodecomposition reaction is quantitative. Experimental confirmation of these two assumptions was possible from a consideration of the material balances. Confidence in the experimental values of R relates directly to the closure of the material balances.

Carbon-labeled reactants can also be used to specify selectivities in a complicated reaction sequence. Consider the reactions to form CO_2 in the sequence of selective hydrocarbon oxidations to form benzene from propylene[6] as in Figure 10 - 5. We used ^{13}C labeled propylene in the presence of unlabeled [^{12}C] hexadiene to determine relative rates of CO_2 from propylene and hexadiene. Pulses of this mixture were reacted at 823°K over Bi_2O_3 in the absence of gaseous O_2 to give the results in Figure 10 - 6. Even though the ratio of $^{12}C/^{13}C$ in the reactant was 5.6/1, the initial ratio of $^{12}C/^{13}C$ in the product CO_2 was 25. Since hexadiene contains twice as many C atoms per molecule as propylene, then the relative rate of hexadiene/propylene combustion is

$$25/[(5.6)(2)] = 2.2$$

These results show that the hexadiene forms carbon dioxide at a rate twice that of propylene.

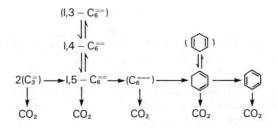

Figure 10 - 5 Reaction Network to Form Carbon Oxides

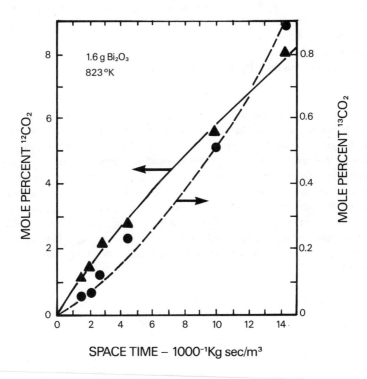

Figure 10 - 6 Distribution of Carbon-13 in Product Carbon Dioxide

Used with permission from M. G. White and J. W. Hightower, *J. Catal.*, **82**, (1983), p. 185.

Isotopically Labeled Oxygen

Isotopic oxygen has been used to study the reaction of gaseous oxygen with oxide catalysts and to study the mechanism of oxygen incorporation into the products of a selective oxidation of hydrocarbons. Novakova[8] and others[9, 10] have reviewed the isotopic exchange (IE) of ^{18}O between the solid and gas phases. We present here only brief excerpts from Reference 8 to introduce the topic and illustrate the applications of IE. One review of propylene selective oxidation[11] includes the ^{18}O incorporation studies that were pivotal to detailing the oxygen incorporation mechanism into the product, acrolein.

Isotopic Exchange of Oxygen

The study of IE is usually conducted concurrently with oxygen incorporation studies for detailing the mechanisms of selective hydrocarbon oxidations. It was learned that incorporation and exchange of oxygen may occur simultaneously in some very few systems and for others the exchange reaction only occurs at temperatures much higher than the temperatures at which the oxidation occurs. For those systems where exchange and reaction occur simultaneously, Boreskov[10] showed definite relationships exist between exchange and oxidation activities. Thus, the results of IE activity may be used to infer the activity of the selective oxidation.

IE Theory

Two cases may be distinguished for IE: (1) heterophase exchange where the solid phase atoms participate in the exchange, and (2) homophase exchange where such participation does not occur. Two types of heterophase exchange may be identified as the result of transient ^{18}O - ^{16}O studies. Single exchange describes only one mechanism by which one oxygen atom from the molecule is exchanged by reaction event with one oxygen atom from the solid. The rate of this exchange is denoted by R'. Multiple exchange describes the replacement of 2 surface oxygen atoms for an oxygen molecule per reaction event. This exchange rate is denoted by R". For the homophase reaction, only one equilibration reaction may be established and denoted by rate, R. It is possible to write equations detailing these three reaction mechanisms.

Single Exchange: R'
$$O^*O^* + O_s \Leftrightarrow O^*O + O_s^* \qquad (10 - 21)$$

$$O^*O + O_s \Leftrightarrow OO + O_s^* \qquad (10 - 22)$$

Multiple Exchange: R"
$$O^*O^* + 2O_s \Leftrightarrow OO + 2O_s^* \qquad (10 - 23)$$

$$O^*O + 2O_s \Leftrightarrow OO + O_s^* + O_s \qquad (10 - 24)$$

$$O^*O + 2O^*_s \Leftrightarrow O^*O^* + O^*_s + O_s \qquad (10 - 25)$$

Homophase Exchange: R $\qquad$ $O^*O^* + OO \Leftrightarrow 2OO^* \qquad (10 - 26)$

The index "s" indicates the surface atom and the asterisk denotes the ^{18}O species. The reaction of a given isotope for the Eqs. (10 - 21 through -26) may be given by the following:

$$d[O_2]/dt = (1/n)Kp_1p_2 \qquad (10 - 27)$$

where

K = R, R', or R"

n = number of molecules converted in the chosen reaction per one molecule of the isotopic species of interest

p_1 = probability of simultaneous presence of molecules and atoms of the given isotopic composition in the chosen reaction

p_2 = probability that a given exchange will lead to a change in the number of molecules of the isotope of interest

Conservation of mass balances may be written for the two oxygen isotopes in the solid and gaseous phases. It is assumed that diffusion in either phases does not influence the observed exchange rate. The derivation of the equation is given in References 8 and 12; only the final equation will be presented here.

$$w - w' = (w_o - w') \exp\{-(2R" + R')[2a + m]t/(2am)\} \qquad (10 - 28)$$

where

w, w_o, w' = number of O^* atoms in gas phase at times $t = t$, $t = 0$, $t = \infty$, respectively

a = total number of molecules (i.e., all isotopes) in gaseous phase

m = total number of atoms (i.e., all isotopes) in solid phase

Equation (10 - 28) gives the total number of ^{18}O atoms in the gas phase which may be distributed between the O^*O^* and O^*O molecules. The expression for the number of O^*O^* molecules in the gas phase, $x(t)$, as a function of time is as follows:

$$x(t) = ac^2/(2a + m)^2 + (w_o - w')[c/(2a + m)] \exp[-(2R" + R')(2a + m)(t)/(2am)] -$$
$$(1/2m)(w_o - w')^2\{[1 + (m^2R/4a^2R") - (mR'/2aR")] \, {}^*$$
$$\exp[-(2R" + R')(2am + m)/(am)(t)]/ \, [2 + (m/2a) - (mR/2aR") + (R'/R")]\} +$$
$$\{x_o - ac^2/(2a + m)^2 - (w_o - w')(c/[2a + m]) + (1/2m)(w_o - w')^2 \, {}^*$$
$$[1 + m^2R/(4a^2R") - (mR'/2aR")]/[2 + (m/2a) - (mR/2aR") + (R'/R")]\} \, {}^*$$
$$\exp[-(1/a)(R + R' + R")t] \qquad (10 - 29)$$

where

c = total number of exchangeable ^{18}O atoms in the system = $w + u$

u = number of exchangeable ^{18}O atoms in the solid phase

The corresponding equation for $y(t)$ is as follows:

$$y(t) = 2ac(2a + m - c)/(2a + m)^2 + (w_o - w')[(2a + m - 2c)/(2a + m)]$$
$$exp[-(2R'' + R')(2a + m)(t)/(2am)] +$$
$$(1/m)(w_o - w')^2\{[1 + (m^2R/4a^2R'') - (mR'/2aR'')]\ *$$
$$exp[-(2R'' + R')(2am + m)(t)/(am)]/\ [2 + (m/2a) - (mR/2aR'') + (R'/R'')]\} +$$
$$\{y_o - 2ac(2a + m - c)/(2a + m)^2 - (w_o - w')[(2a + m - 2c)/(2a + m)]- (1/m)(w_o - w')^2\ *$$
$$[1 + m^2R/(4a^2R'') - (mR'/2aR'')]/[2 + (m/2a) - (mR/2aR'') + (R'/R'')]\}\ *$$
$$exp[-(1/a)(R + R' + R'')t] \qquad (10 - 30)$$

At the beginning of an IE experiment the variable, a, may be calculated knowing the temperature, pressure, and volume of the system, and variable, m, knowing the weight and constitution of the solid. The variable, w, may be calculated from the mass spectral analysis of the gas phase as a function of time. The data of $(w - w')$ may be fit to a semilog plot to determine the parameter $(2R'' + R')$; however, the unique values of R' and R'' cannot be determined by this single measurement. From the same data, the variable, $x(t)$, is used together with the value of c in to determine the parameters R', R'' if $R = 0$. Clearly, c will be invariant with time if all the solid phase oxygen atoms are exchangeable.

Estimates of R, R', and R'' are difficult to calculate if none can be assumed to be zero initially. If all mechanisms make significant contributions to the total exchange process, then the data of $w(t)$ (total O^* atoms in gas phase), $x(t)$ (number of O^*O^* molecules in gas phase), and $y(t)$ (number of O^*O molecules), or $z(t)$ (number of OO molecules) must be fit to nonlinear equations such as Eq. (10 - 29, - 30) to extract the parameters R, R', and R''. It may be possible to eliminate one or more of the mechanisms by considering the data of w, x, y, and z as functions of time. If $2x + y$ remains invariant with time (i.e., the gas phase concentration of O^* atoms doesn't change), then the R' and R'' are essentially zero and only homophase exchange is occurring. In this instance we are not considering the trivial case where the time derivatives of x, y, and z are all zero.

$$x'(t) = y'(t) = z'(t) = 0$$

On the other hand, should the sum of the variables $(2x + y)$ decrease with time, then R' and R'' are significant. The relative magnitudes of R' and R'' may be estimated by another manipulation of the data for a gaseous mixture initially at isotopic equilibrium. If the function $(y^2)/(xz)$ attains a minimum with time, then the R'' mechanism controls the exchange process. Conversely, if $(y^2)/(xz)$ remains essentially constant and equal to 4, then the R' mechanism dominates the exchange process.

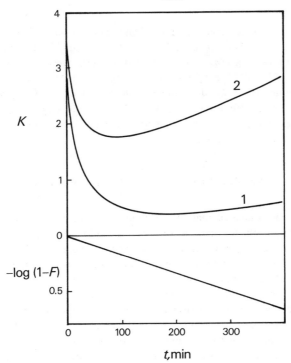

Figure 10 - 7 Labeled Oxygen Content for Scrambling over ZnO

Top: The time dependence of K for ZnO, $K = y^2/xz$: circles, experimental data. Solid Lines: 1, theoretical curve for the limiting case $R = R' = 0$; 2, theoretical curve fitted by the computer with exchange rates $R = 0.34$, $R' = 0.24$, and $R'' = 0.14 \times 10^{17}$ molecules/min-m². Bottom: The time dependence of -log(1-F), $F = (w-w')/(w-w_o)$; ZnO; commercial, AR grade, surface area 2.7 m²/g, temperature = 485 C, pressure of oxygen = 0.5 Torr. Used with permission from J. Novakova, P. Jiru, and K. Klier, *Collect. Czech. Chem. Comm.,* **33**, (1968), p. 1338.

These preliminary considerations suggest several experiments to determine the dominant exchange mechanism and to estimate the governing parameter(s). First, a noncatalytic exchange experiment is run to evaluate the size of the homophase exchange rate, R, and in some cases to establish those conditions which minimize R. If R can be made small, then Eq. (10 - 29) may be simplified somewhat. If R is finite at the desired reaction conditions, then the initial gas mixture must be allowed to equilibrate before it is contacted with the solid. The second experiment would begin by contacting the equilibrated gas over a solid devoid of O^* atoms, except for natural abundance. The response of the function $(y^2)/(xz)$ with time would indicate if both of the heterophase mechanisms are operative. Should only one of the heterophase mechanisms dominate (i.e., $R = 0$ and either R' or $R'' = 0$), then the rate mechanism can be found using Eq. (10 - 28) for the variable $(w - w')$. However, should two or three of the mechanisms be finite and significant, a modified gradient method for computer programming may be employed.[13] Figure 10 - 7 shows the exchange data for a commercial grade (AR) ZnO at 485°C and oxygen parital pressure equal to 0.5 Torr. The surface area of the zinc oxide was 2.7 m^2/g. The semilog plot of $(w - w')/(w_o - w')$ (denoted F here) versus time in minutes is linear. Clearly, R' and R" are finite and significant. The function $K = (y^2)/(xz)$ also shows a minimum (curve 2) which means that R' and R" are finite but the data do not describe the same curve for which $R = R' = 0$ (curve 1). Thus, a computer fit was used to describe the data. The best fit employed these values of R, R', and R": 0.34, 0.24, and 0.14 x 10^{17} molecules/min-m^2.[14]

Relation of Exchange Rates to Catalyst Activity

It has been speculated that the reaction of oxygen with the catalyst surface is the rate limiting step for certain selective oxidations; for example, the rate of reduction of Cu$_2$O by propylene occurs very rapidly at 200°C; whereas the reoxidation of the catalyst with pure oxygen does not occur for temperatures less than 300°C[15] (Fig. 10 - 8). These data suggest that the rate of the selective oxidation reaction of the catalyst may be limited by the rate of catalyst reoxidation. The nonintegral reaction orders with respect to oxygen[16] suggest that the oxygen molecule is activated by the surface. These findings may be illuminated by the results of isotopic oxgyen studies. Activation of oxygen should be apparent by the magnitude of R' next to the other mechanisms. Moreover, if the rate limiting step is related to reaction of oxygen with the catalyst, then the exchange rates should be equal to the reaction rates.

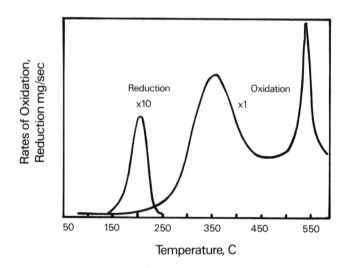

Figure 10 - 8   Rates of Oxidation and Reduction of Cuprous Oxide

Consider the findings of Antoshin et al.[17] who ruled out activation of O_2 as the rate limiting step in the oxidation of hydrogen over rare earth oxides. Only homomolecular exchange rates of oxygen were observed and these rates were one-fifth that of the oxidation rates. Moreover, the first-order kinetics reinforced these findings that an Eley-Rideal mechanism was operative. Finally, the reactivity of the catalysts paralleled the rates of hydrogen-deuterium exchange, suggesting that activation of hydrogen was the rate limiting step.

Comparison of absolute reaction rates is not as enlightening as comparison of trends over a series of catalysts. Consider the rates of hydrogen oxidation and isotopic exchange rates (Fig. 10 - 9) over the Lanthanide series of rare earth oxides.[18] There is agreement between the activity of the oxidations and isotopic exchange rates, suggesting the rate limiting step for each was the same. These data suggest that agents incorporated into the catalyst to increase the IE reaction rate will also increase the reactivity of the catalyst towards oxidation of hydrogen.

These few examples show the relationship between the IE reaction and the reactivity of the catalyst; however, there is a wealth of data to show the opposite is true. There are ample data to show that catalyst oxygen does not participate in the reaction[10] or the reactive sites represent only a small fraction of the surface.[19] Finally, the reaction temperature of some systems is much lower than the temperatures required to sustain IE over the same catalysts.[20]

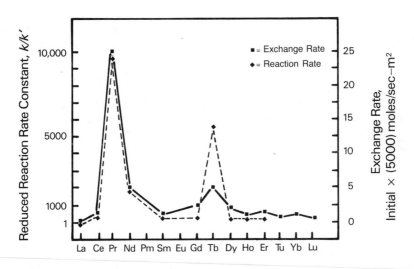

Figure 10 - 9 Rates of Isotopic Oxygen Exchange and Hydrogen Oxidation

Used with permission from Kh. M. Minachev, *Catalysis*, Vol I, *5th Int. Cong. Catal.* (Amsterdam: North-Holland Publ. Co., 1974), p. 219.

Oxygen Incorporation

The role of oxygen in the selective oxidation reaction has been studied extensively. Oxygen incorporation into the oxide and into the oxygenated products may be viewed as a stepwise process beginning with the adsorption and activation of oxygen onto the catalyst surface.

$$O_2(g) \rightarrow O_2(ads) \rightarrow O_2 \rightarrow 2O \rightarrow 2O^= \qquad (10 \cdot 31)$$

At any point in this sequence, the oxygen species may be incorporated into the product(s) of interest according to the appropriate mechanism. The use of ^{18}O tracers has led to a better understanding of these mechanisms.

Examples in the literature show the oxygen incorporation into the oxygenated products may arise almost exclusively from the lattice oxygen contained in the solid phase. These conclusions were based on the results of several ^{18}O studies involving the selective oxidation of propylene over bismuth molybdate. One type of experiment involved the oxygen exchange with the solid at temperatures between $250°$ and $500°C$. No change in the isotopic composition of the gas phase over a two-hour period was observed nor were there any scrambling of the isotopes in the gas phase. Second, a propylene/^{18}O feed produced acrolein and carbon dioxide containing no more than 2.5% ^{18}O.[11] These results demonstrate that catalyst oxygen was incorporated into the products.

The use of isotopically labeled oxygen is not restricted to molecular oxygen. It has been used in the form of the labeled catalyst to confirm the participation of lattice oxygen in a particular mechanism. However, several caveats are appropriate. It is very expensive and difficult to prepare a reasonably sized catalyst sample having all the oxygens labeled. Thus, only the surface oxygens may be labeled during an exchange process with a ^{18}O containing gas. If the catalyst is heated subsequently to this surface enrichment, rapid diffusion of the lattice oxygen ions into the bulk may destroy the surface enrichment. Another mechanism to destroy the surface enrichment of the ^{18}O may occur by isotope scrambling with another oxygen containing species such as water, CO, CO_2, and so on.

Isotopically labeled water has been used to study the selectivity of propylene oxidation over a tin molybdate catalyst at $365°C$. Here the labeled water was introduced as a co-feed to the propylene and unlabeled oxygen. The side product, acetone, was observed to be the only product to contain the labeled oxygen indicating the hydration of an intermediate (perhaps the allyl radical) led to the formation of the acetone while no carbon oxides were derived from oxidation of the acetone.

Conclusions

This brief overview of labeled compounds and their applications was intended to motivate readers towards devising their own experiments. The techniques are straightforward, requiring the appropriate detection device and the same type of chemical reactor as may used in collecting the kinetics. The author prefers the all-glass, recirculation reactor to collect the reaction kinetics, perform the labeled compound studies, and if the system is attached to the appropriate volumetric device, then BET surface areas and selective chemisorptions may also be performed in the same device. The beauty in using labeled compounds is that some useful information almost always arises from the results.

REFERENCES

1. White, M. G. and J. W. Hightower, *Am. Inst. Chem. Eng. J.,* **27**, (1981), p. 535.

2a. Melander, Lars, *Isotope Effects on Reaction Rates* (New York: Ronald Press, 1960).

2b. Edgell, W. F. & C. J. Ultee, *J. Chem. Phys,* **72**, (1954), p. 1983.

3a. Adams, C. R. and T. J. Jennings, *J. Catal,* **2**, (1963), p. 63.

3b. Adams, C. R. and T. J. Jennings, *J. Catal.,* **3**, (1964), p. 549.

4. Adams, C. R., *J. Catal.,* **10**, (1968), p. 355.

5. Adams, C. R., *J. Catal.,* **11**, (1968), p. 96.

6. White, M. G. and J. W. Hightower, *J. Catal.,* **82**, (1983), p. 185.

7. Sachtler, W. M. H. and N. H. DeBoer, *3rd Inter. Cong. Catal.* (Amsterdam: North-Holland, Publ., Inc., 1964), Vol. I, p. 252.

8. Novakova, J., *Catal. Rev.,* **4**, (1970), p. 77.

9. Winter, E. R. S., *Adv. Catal.,* **10**, (1958), p. 196.

10. Boreskov, G. K., *Adv. Catal.,* **15**, (1964), p. 285.

11. Keulks, G. W., L. D. Krenzke, and T. M. Noterman, Adv. Catal. , **27**, (1978), p. 183.

12. Klier, K., J. Novakova, and P. Jiru, *J. Catal.,* **2**, (1963), p. 479.

13. Muzykantov, V. S., P. Jiru, K. Klier, and J. Novakova, *Collect. Czech. Chem. Comm.,* **33**, (1968), p. 829.

14. Novakova, J., P. Jiru, and K. Klier, *Collect. Czech. Chem. Comm.,* **33**, (1968), p. 1338.

15. White, M. G., private notes (1982).

16. Mamedov, E. A., E. G. Gamid-zade, F. Agaev, and R. G. Rizaev, *Kinet. Katal.,* **20**, (1979), p. 410.

17. Antoshin, G. V., Kh. M. Minachev, and M. E. Lokhuary, *J. Catal.,* **22**, (1971), p. 1.

18. Minachev, Kh. M., *Catalysis,* Ed. by J. W. Hightower,. Vol. I (Amsterdam: North Holland/Elsevier, 1974), p. 219.

19. Winter, E. R. S., *J. Chem. Soc.,* (1955), p. 3824.

20. Margolis, L. I., *Geterognnoe Kataliticherskoeokislenie Uglevodorodov Izd. Khimiia*(Leningradskoe Otd.: 1967).

21. Sagun, G., Ph. D. thesis, Georgia Institute of Technology (1982).

PROBLEMS

1. The statistical distribution of deuterium atoms in a molecule can be predicted knowing the initial ratio of deuterium/hydrogen in the reaction mixture and the number of exchangeable hydrogens in the molecule. For example, benzene has six exchangeable hydrogens and if we react an equilimolar mixture of perdeuterated and undeuterated benzene, the initial D/H is unity. The distribution of deuterium atoms (d_i) for equilibrium exchange is the combinatorial distribution (c_i) weighted by the D/H ratio as in the following:

$$d_i = c_i(D/H)^{i+1}/\Sigma c_i(D/H)^{i+1} \qquad (10 - 32)$$

Find the d_i for D/H = 0.5, 0.921, 1, and 2. Plot these distributions as bar graphs for i = 0 to 6.

2. One may calculate an extent of conversion for a deuterium exchange by summing the observed distribution of deuteriums incorporated into the product.

$$\phi = \Sigma\, i\, d_i \qquad (10 - 33)$$

Consider the D/H exchange of benzene and deuterium gas over a Pt/alumina catalyst. The exchange is modeled by first-order kinetics such that the integrated rate expression is

$$ln(\phi\infty - \phi) = ln(\phi\infty) - kt/\phi\infty \qquad (10 - 34)$$

Using the data of Sagun[21] (Table 10 - 1), determine k.

Table 10 - 1 Deuterium Incorporation in Benzene as a Function of Time for the Exchange over Pt/Alumina

t, min	d_0	d_1	d_2	d_3	d_4	d_5	d_6
0							
3	0.8557	0.1330	0	0.0058	0	0	0.0002
6	0.7846	0.1509	0	0.0196	0.0027	0.0056	0.0066
11	0.6743	0.2381	0.0416	0.0307	0.0068	0.0055	0.0032
18	0.5778	0.2825	0.0814	0.0421	0.0111	0.0040	0.0015
27	0.4382	0.3038	0.1362	0.0651	0.0287	0.0231	0.0046
35	0.3698	0.3078	0.1650	0.0942	0.0415	0.0265	0.0051
45	0.2849	0.3026	0.1977	0.1216	0.0607	0.0239	0.0086
55	0.2489	0.2885	0.2163	0.1355	0.0710	0.0289	0.0130

Used with permission from G. Sagun, Ph. D. thesis, Georgia Institute of Technology, Atlanta, GA, (1982).

INDEX